Drop, Bubble and Particle Dynamics in Complex Fluids

Drop, Bubble and Particle Dynamics in Complex Fluids

Special Issue Editors

Pengtao Yue

Shahriar Afkhami

MDPI • Basel • Beijing • Wuhan • Barcelona • Belgrade • Manchester • Tokyo • Cluj • Tianjin

Special Issue Editors
Pengtao Yue
Virginia Tech
USA

Shahriar Afkhami
New Jersey Institute of Technology
USA

Editorial Office
MDPI
St. Alban-Anlage 66
4052 Basel, Switzerland

This is a reprint of articles from the Special Issue published online in the open access journal *Fluids* (ISSN 2311-5521) (available at: https://www.mdpi.com/journal/fluids/special_issues/drop_bubble_particle_dynamics).

For citation purposes, cite each article independently as indicated on the article page online and as indicated below:

LastName, A.A.; LastName, B.B.; LastName, C.C. Article Title. *Journal Name* **Year**, *Article Number*, Page Range.

ISBN 978-3-03928-296-8 (Pbk)
ISBN 978-3-03928-297-5 (PDF)

Contents

About the Special Issue Editors

Pengtao Yue is an Associate Professor in the Department of Mathematics at Virginia Polytechnic Institute and State University. He received his Ph.D. in fluid mechanics from the University of Science and Technology of China. His current research focuses on the numerical simulation of complex fluids and moving boundary problems, such as viscoelastic fluids, particulate flows, drop dynamics, wetting dynamics, and phase transition.

Shahriar Afkhami is an Associate Professor in the Department of Mathematical Sciences at the New Jersey Institute of Technology. He received a Ph.D. from the Mechanical and Industrial Engineering Department at the University of Toronto, Canada. He is interested in the computational and mathematical modeling of complex systems including viscoelastic liquids, electro/ferrohydrodynamics, interfacial flows in porous media, and micro/nanofluidics, as well as in applications of high-performance computing.

fluids

Editorial

Editorial for Special Issue "Drop, Bubble and Particle Dynamics in Complex Fluids"

Shahriar Afkhami [1,*] and Pengtao Yue [2,*]

1 Department of Mathematical Sciences, New Jersey Institute of Technology, Newark, NJ 07102, USA
2 Department of Mathematics, Virginia Tech, Blacksburg, VA 24061, USA
* Correspondence: shahriar.afkhami@njit.edu (S.A.); ptyue@math.vt.edu (P.Y.)

Received: 20 December 2019; Accepted: 26 December 2019; Published: 2 January 2020

The presence of drops, bubbles, and particles affects the behavior and response of complex multiphase fluids. In many applications, these complex fluids have more than one non-Newtonian component, e.g., polymer melts, liquid crystals, and blood plasma. In fact, most fluids exhibit non-Newtonian behaviors, such as yield stress, viscoelastity, viscoplasticity, shear thinning, or shear thickening, under certain flow conditions. Even in the complex fluids composed of Newtonian components, the coupling between different components and the evolution of internal boundaries often lead to complex rheology. Thus, the dynamics of drops, bubbles, and particles in both Newtonian fluids and non-Newtonian fluids are crucial to the understanding of the macroscopic behavior of complex fluids. The goal of this Special Issue was to gather recent experimental, numerical, and theoretical research on drop, bubble, and particle dynamics in complex fluids.

Leiva and Geffory [1] experimentally investigated the variation of droplet size distribution of emulsions under slow shearing flow. In contrast to the good stability of emulsions at rest, the size distribution changes significantly due to breakup and coalescence of droplets under flow. A bimodal size distribution and a banded structure were observed at lower and higher shear rates, respectively.

Amah et al. [2] performed direct numerical simulations on the motion of dielectric particles in electric fields of microfluidic devices, where the rigid-body motion of particles was enforced by a distributed Lagrange multiplier method and the electric force acting on the particles was computed using the point-dipole and Maxwell stress tensor approaches. Their numerical results revealed that the tendency of particles to form chains diminishes when the particle size is comparable to the spacing between electrodes, due to the modification of the electric field by the presence of particles.

Poryles and Zenit [3] experimentally studied the rising of Newtonian oil drops in a shear-thinning viscoelastic liquid. A so-called rising velocity discontinuity was observed for drops larger than a certain size. Beyond the critical velocity, the drop forms a long tail, which emits small droplets. The size and emission frequency of the droplets were found to be dependent on the volume of the mother drop. Potentially, this setup can be used to generate small droplets with desirable sizes by adjusting the volume of the rising drop.

Spanjaards et al. [4] numerically investigated the migration of sedimenting particles in a viscoelastic Couette flow between two rotating cylinders. An arbitrary-Lagrangian–Eulerian moving-mesh method was used to track the moving particles, and the DEVSS-G and log-conformation representation were used for the viscoelastic stress. The migration velocity of a sedimenting particle in a Couette flow was found to be higher than the sum of migration velocities due to sedimentation and Couette flow individually.

Hamidouche et al. [5] investigated the performance of a discrete element method (DEM)/large-eddy simulation (LES) solver for the prediction of gas-particle flows in a fluidized bed. Mesh sensitivity, wall conditions for the gas phase, and particle-wall and particle-particle friction coefficients were systematically studied. Good agreements with experimental data were achieved if the numerical parameters were properly chosen.

Helmers et al. [6] developed a mathematical model to predict the excess velocity of Taylor drops in square microchannels. The proposed model was adapted with a stochastic and metaheuristic optimization approach based on genetic algorithms and compared well with high-speed camera measurements and published empirical data.

Zaleski and Afkhami [7] analyzed the dynamics of ellipse-shaped droplets, either conducting or dielectric, in an electric field using conformal maps. Different from previous analytical work in the literature, the complexity of boundary conditions at the electrode was also considered. In the conducting case, the maximum droplet height is attained when the distance between the electrode and the drop becomes sufficiently large; in the dielectric case, hysteresis can occur for certain values of electrode separation and relative permittivity.

Funding: This research received no external funding.

Conflicts of Interest: The authors declare no conflict of interest.

References

1. Leiva, J.; Geffroy, E. Evolution of the Size Distribution of an Emulsion under a Simple Shear Flow. *Fluids* **2018**, *3*, 46. [CrossRef]
2. Amah, E.; Janjua, M.; Singh, P. Direct Numerical Simulation of Particles in Spatially Varying Electric Fields. *Fluids* **2018**, *3*, 52. [CrossRef]
3. Poryles, R.; Zenit, R. Encapsulation of Droplets Using Cusp Formation behind a Drop Rising in a Non-Newtonian Fluid. *Fluids* **2018**, *3*, 54. [CrossRef]
4. Spanjaards, M.; Jaensson, N.; Hulsen, M.; Anderson, P. A Numerical Study of Particle Migration and Sedimentation in Viscoelastic Couette Flow. *Fluids* **2019**, *4*, 25. [CrossRef]
5. Hamidouche, Z.; Dufresne, Y.; Pierson, J.; Brahem, R.; Lartigue, G.; Moureau, V. DEM/CFD Simulations of a Pseudo-2D Fluidized Bed: Comparison with Experiments. *Fluids* **2019**, *4*, 51. [CrossRef]
6. Helmers, T.; Kemper, P.; Thöming, J.; Mießner, U. Modeling the Excess Velocity of Low-Viscous Taylor Droplets in Square Microchannels. *Fluids* **2019**, *4*, 162. [CrossRef]
7. Zaleski, P.; Afkhami, S. Dynamics of an Ellipse-Shaped Meniscus on a Substrate-Supported Drop under an Electric Field. *Fluids* **2019**, *4*, 200. [CrossRef]

fluids

Article

Evolution of the Size Distribution of an Emulsion under a Simple Shear Flow

Jairo M. Leiva * and Enrique Geffroy

Instituto de Investigaciones en Materiales, Universidad Nacional Autónoma de México, Ciudad Universitaria, 04510 Cd. de México, CDMX, Mexico; geffroy@unam.mx

* Correspondence: eleiva@iim.unam.mx; Tel.: +52-55-5622-4632

Received: 25 April 2018; Accepted: 20 June 2018; Published: 25 June 2018

Abstract: Understanding the rheology of immiscible liquids mixtures, as well as the role played by its micro-structures are important criteria for the production of new materials and processes in industry. Here, we study changes over time of the droplet size distributions of emulsions induced by slow shearing flows. We observe that the initial heterogeneous microstructure may evolve toward more complex structures (such as bimodal distribution) as a result of coalescence and rupture of droplets. These dynamic structures were produced using a flow cell made up of two parallel disks, separated by a gap of 100 μm. The steady rotation of the lower disk generates a simple shear flow of $\dot{\gamma} = 0.75\ s^{-1}$, during $\sim$ 400 s. After a brief rest time, this procedure was repeated by applying a step ramp until the maximum shear rate of $4.5\ s^{-1}$ was reached, using step increments of $0.75\ s^{-1}$. During the last portion of the flow and during the rest time in between flows, structures of emulsions were characterized. Initially, a broad single-peak distribution of drops was observed, which evolved toward a rather narrower bimodal distribution, at first due to the coalescence of the smaller droplets and subsequently of the larger drops. The rupture of drops at higher shear rates was also observed. The observed evolutions also presented global structures such as "pearl necklaces" or "bands of particles", the latter characterized by alternating bands of a high density of particles and regions of the continuous phase with only a few droplets. These changes may indicate complex, time-dependent rheological properties of these mixtures.

Keywords: emulsion microstructure; drop size distribution; monomodal–bimodal distributions

1. Introduction

Emulsions are of great relevance for a variety of applications in food, pharmaceuticals, adhesives, cosmetics, plastics, fertilizers, and petroleum recovery industries, inasmuch as mixtures of fluids provide a wider range of properties in their final products. However, the properties of the mixtures depend, to a great extent, on the microstructure of the emulsion, which is, in turn, the result of the history of the flow. Therefore, most rheological properties must be considered as dynamical properties, which may depend on the structure of the emulsion in a nonlinear manner.

For emulsions, the time evolution of their rheological properties can depend on a number of factors, such as fluid properties—i.e., the fluid viscosities and the interfacial energy—the particle size and shape, volume fractions of the phases, and other subtler effects, such as the anisotropic distribution of particles that the imposed flow can induce [1]. In the literature, we find studies focused on the effects characterized by important processes, such as the role of coalescence-rupture [2–4] of drops or parameters (e.g., the viscosity ratio [5,6], the shear rate [7–9], and the volume fraction [10,11]). Nevertheless, much remains to be explored mainly because a complete understanding of the observed phenomena [12] is still lacking. This is especially relevant when the structure evolution shows multiple well-defined structures as time passes under a flow field, particularly, when *global anisotropic structural changes* are induced at higher shear rates. Despite the possible relevance of the latter, here, we address the changes in the microstructure of the emulsion at low shear rates and in the relevance to the rheology of the emulsion. This behavior corresponds to the low, dimensionless time evolutions induced by slow shearing flows presented in this work.

1.1. Theoretical Background

There are many results published on the effects of the history of the shear rate (e.g., on the deformation, breakup of a single drop, or coalescence of quasi-equal size drops). The pioneering work of G. I. Taylor [13] about the slight deformation of a single drop establishes experimentally that two dimensionless numbers mainly determine the drop deformation under a linear flow: the capillary number and the viscosity ratio. Larger capillary numbers induce larger drop deformations, and viscous drops require a higher shear rate or capillary number in order to deform significantly.

The capillary number is given by Equation (1)

$$Ca = \frac{\eta_m R \dot{\gamma}}{\sigma}, \tag{1}$$

where η_m, R, $\dot{\gamma}$, and σ represent the viscosity of the fluid matrix, the radius of the drop, the applied shear rate, and the interfacial tension coefficient between phases, respectively.

The viscosity ratio between the two phases is given by Equation (2)

$$p = \frac{\eta_d}{\eta_m}, \tag{2}$$

where η_d is the viscosity of the disperse fluid, and η_m is the viscosity of the continuum fluid or matrix.

When studying emulsions under flow (besides deformation of drops), other phenomena can frequently be observed, such as coalescence of drops [4], break up of drops [6], or capture of a rather small drop by another significantly larger drop [14]. When applying a larger shear rate to the emulsion, these phenomena may be observed and are frequently the main source of the observed dynamical changes of the size distributions of drops; although, each phenomenon depends most likely on different physical mechanisms. Under weak flows, coalescence of small, equal-sized drops is observed, and the rupture of drops occurs at higher rates of deformation, especially with low viscosity fluids. Research reporting large induced deformations under shearing flows, even beyond a critical drop size (up to break-up into two or multiple droplets), are given in [15–17]. The growth of drops through the capture of much smaller, nearby droplets—by a mechanism that appears to resemble an Oswald ripening process—can be readily produced but requires a rather broad size distribution, including large drops. Finally, spatially varying distributions of particles induced by flow have been observed as well, but its relation to the former phenomena is less well documented. Here, we attempt to describe a robust technique to evaluate the slow-shear-flow phenomena, which generally modifies the observed rheology of an emulsion in a rather complex and nonlinear manner.

1.1.1. Coalescence in Slow Flows

Here, the coalescence mechanism occurs mainly in the weakest of flows. During this process, two drops may coalesce if they spend enough time in close proximity. Thus, a dimensionless time, τ, indicative of the minimum time required for a high probability of coalescence (or its efficiency) can be calculated as

$$\tau = \dot{\gamma} t_e, \tag{3}$$

where t_e corresponds to the duration of imposed flow of the experiment, while the shear rate, $\dot{\gamma}$, is *proportional* to the rate of collisions of drops of a given size. Please note that it is customary to consider Equation (3) as a deformation measure, but here, we prefer to associate the inverse of this dimensionless number to an efficacy or efficiency of coalescence. Dimensionless deformation measures are most appropriate when studying fluids with a homogeneous and continuous micro-structure, such as polymer solutions (with an associated characteristic time-scale spectrum), etc. In contrast, for our experiments, a dimensionless time is more closely associated with inverse frequency of events. This may provide a better understanding between the statistics of the drops distributions and the experimental conditions. The inverse time for shear rate is associated here with a value proportional to the frequency of collision of drops and is not relevant as a measure of rate of deformation.

This interpretation allows us to compare different flow regimes, assuming that other present phenomena remain stationary. The proportionality constant for the rate of collisions is a rather complex function of the hydrodynamics of multiple interacting drops. For isolated pairs of drops (i.e., very dilute emulsions), the film drainage model is frequently used, which is more suited for non-deformable surfaces (i.e., drops of high viscosity) [18,19].

For emulsions, mean field values for the drainage model are difficult to calculate. In contrast, when evaluating the evolution of the size distributions under flow, coalescence between the smallest of drops can be inferred rather easily, because the rate of decrease of frequency for the smallest drops is about twice the rate of increase of frequency of drops with double their volume. The same is true for the larger-size drops of multimodal distribution, as will be shown subsequently. Thus, the coalescence rate can be established by the product of the frequency of the drops' collision times the efficiency of coalescence [20].

For emulsions with a high fraction of the disperse phase—characterized by a broad size distribution and under weak shearing flows—the experimental information is difficult to interpret, due mainly to a high density of very small drops, especially at the onset of the flow. It is also evident that an upper limit to the drop size (spheroidal drops and for a given shear rate) exists when the kinetics of the rupture of the drops competes with the coalescence phenomenon, and the changes of the distribution of small drops vanish almost completely [21]. Given that the capillary number increases for the larger drops, then a maximum size exists, where coalescence dominates and rupture kinetics begin, cancelling each other out. In the work of Grizzuti and Minale, it is suggested that the two processes coexist in the same system [19–27]. Therefore, a second critical capillary number should be observed—associated with the transition of coalescence to rupture—and defined when drops increasing in size undergo a breakage process [11,12,23,24]. The purpose of this work is to clarify the presence of coalescence and rupture processes under shearing flows by studying the evolution of the drop size distribution of the dispersed phase.

1.1.2. Breakup of Droplets under a Shearing Flow

The so-called critical capillary number required for the rupture of a vesicle, Ca_{rup}, appears to depend principally on p for simple shear flows, as shown by Grace [16] and De Bruijn [17]. A broad set of data for drops deformation and break up, including a large class of two-dimensional (2D)-flows—covering from simple shear up to a purely elongational flow—was provided by Bentley and Leal [28]. More recently, for emulsions subjected to simple shear flows, Jansen [29] has shown that the critical capillary number decreases with increments of the fraction of the disperse phase: $Ca_{rup}(p, \phi)$. Droplet-breaking mechanisms and shapes of Newtonian liquid droplets have been extensively studied. If $Ca << 1$, the drop shape is slightly ellipsoidal, depending on p, and aligned at an orientation angle of 45° with respect to the direction of flow. As the capillary number increases, the steady state elongation grows, and the drop rotates aligning itself along the direction of flow. For higher capillary numbers, beyond the critical value, rupture is observed with the breaking mode depending on the viscosity ratio. For $p < 1$, the drops assume an elongated highly cusped form, from which small drops (the so-called tip streaming phenomenon) are launched. For p approximately equal to 1, the central portion of the droplet forms a neck (or necks) followed by the breaking up into two daughter-droplets, with small satellite droplets between them. $Ca >> Ca_{rup}$, droplets are deformed into long, thin filaments that eventually break up through the instability of the capillary wave mechanism. These mechanisms become more complex as the density of disperse phase drops increases.

2. Materials and Methods

2.1. Constituents and Preparation of Emulsions

Two immiscible fluids were prepared as the emulsion disperse-continuum components, looking for a pair of liquids with high viscosities and equal densities: an aqueous solution, as the dispersed phase, and a mixture of alkanes. The aqueous solutions is (w/W) 10 μM polyethylene oxide (with a viscosity-averaged molecular weight of $M_v \sim$ 1,000,000 , Sigma-Aldrich CAS#372781, Sigma-Aldrich, St. Louis, MO, USA) in 97% ultra-pure water (resistivity ≥18.2 MΩ·cm; ρ = 0.997 g/mL)) and 3% 2-propanol (Sigma-Aldrich CAS#190764, ≥99.5% Reagent grade). The continuum phase is a mixture of eicosane (Sigma-Aldrich CAS#219274), heptadecane (Sigma-Aldrich CAS#128503), 1,2,4-trichlorobenzene (Sigma-Aldrich CAS#132047), and polybutadiene (Sigma-Aldrich CAS#181382). The alkane fluid is prepared by first mixing 7.56% eicosane with 39.69% heptadecane in a glass bottle, while maintaining it at 30 °C, and then adding 46.5% 1,2,4-trichlorobenzene and 6.25% polybutadiene ($M_n \sim$ 200,000). The viscosities of the fluids were measured with an ARES G2 Rheometer (TA Instruments, New Castle, DE, USA) using the concentric cylinder geometry. The aqueous phase has a viscosity of 0.57 Pa·s and the continuum phase a viscosity of 2.08 Pa·s at 30 °C; the viscosity ratio is $p = 0.27$. The densities (g/cm^3, at 30 °C) are 0.98 and 0.95 for the aqueous phase and the continuum phase, respectively. By using 2-propanol (aqueous solution) and trichlorobenzene (alkane mixture), the density of the two fluids can be adjusted to minimize sedimentation in the emulsion. The interfacial tension σ was determined by the deformed drop retraction (DDR) method, as described by Guido and Villone [30], using the optical shear cell CSS450, which is the same cell that we use in this work. The measured averaged surface tension is 0.11 mN/m, evaluated for a set of 11 drops of aqueous fluid (40%) in the oil phase.

The emulsion was prepared by mixing a 50 wt % oil phase with 50 wt % water phase using an homogenizer (Omni Inc., Kennesaw Georgia, GA, USA) with a generator probe 10 mm × 95 mm fine saw tooth (SKU#15051), spinning at 3000 rpm during 300 s, and at a constant temperature of 30 °C. Afterwards, the emulsion was placed in glass tubes (5 mm inner diameter) at rest for 48 h and at 30 °C. In this way, the stability of the emulsion was visually evaluated after 48 h, ensuring that trapped air bubbles were removed. The sample of the emulsion was placed on the bottom plate of the flow cell, and the top plate was carefully placed on top, while the plates were being slowly compressed (squeezed) to reduce the separation until reaching a gap of 100 μm. This compression process reduces residual stresses while maintaining the homogeneity of the sample. Following loading, the cell was allowed to relax for a period of ~600 s. All flow experiments started with an initial pre-conditioning of the emulsion to eliminate possible residual stresses by subjecting it to a shear rate of 0.075 s^{-1} for ∼500 s. During this initial flow, the drop size data was dominated by a large count of very small drops—with diameters below the resolution of the optical arrangement ≤5 μm—and no comparison with the drop distribution after the loading of the cell is considered reliable. At this instant, we set $\tau \equiv 0$.

Subsequently, a ramp sequence of constant steady shear flows was applied for each sample. Each step of the ramp consisted of a steady flow applied during ∼400 s, followed by a no-flow rest time of ∼18 s, sufficiently long for drops to attain a spherical shape. The initial step of the ramp began with a flow of $\dot{\gamma} =$ 0.75 s^{-1}, and subsequent flows stepped up by increments of 0.75 s^{-1} up to 4.5 s^{-1}. Toward the end of each steady shear stress—just before the flow was stopped—and a short time after the no flow condition prevailed, a set of images was taken for the statistical analysis. For all the experiments, the no-flow rest time appears to have a negligible effect on the drop size distributions while facilitating the determination of the size of particles of a quasi-spherical shape. Thus, this multistep history of flow stresses is responsible for the changes in the distribution of drops in the sample. The influence of stopping the flow upon the global dimensionless time $(\tau = \dot{\gamma} t_e)$, as well as the structure of the distribution, can be considered negligible. Here, no inertia effects on the deformation of drops is expected given that the nominal Reynolds number—based on tangential velocity, cell gap, and viscosity of the continuum phase—are within 3×10^{-6}–2×10^{-5}, for all shear rates.

2.2. Smooth Kernel Distribution Estimation

The smoothing of all drop size frequency histograms was done with the kernel density technique in order to obtain the probability density function with a known collection of frequency points. Here, the histogram area under the curve is assumed to be 1, and the probability of a drop diameter d_i corresponds to the area under the curve between those two points $(d_i,\ d_i + \Delta d)$, where Δd is the difference between diameters [31]. This tool provides a quick evaluation of the distribution as a continuous function; the smoothing parameter (*bandwidth)* used in all histograms is 1.25 [32]. We also evaluated the quality of polydispersity for the drop distributions via a polydispersity index based on the average drop size (diameter) $D_{1,0}$, the average volume size $D_{4,3}$, and the contributions of the tails of the distribution, respectively:

$$d_N = D_{1,0} = \frac{\sum_{i=1}^{\infty} n_i \cdot d_i}{\sum_{i=1}^{\infty} n_i}, \quad (4)$$

$$d_V = D_{4,3} = \frac{\sum_{i=1}^{\infty} n_i \cdot d_i^4}{\sum_{i=1}^{\infty} n_i \cdot d_i^3}, \quad (5)$$

$$skewness = \frac{n}{(n-1)(n-2)} \sum_{i=1}^{n} \left(\frac{d_i - \bar{d}}{sd} \right)^3, \quad (6)$$

$$kurtosis = \frac{n(n+1)}{(n-1)(n-2)(n-3)} \sum_{i=1}^{n} \left(\frac{d_i - \bar{d}}{sd} \right)^4 - 3\frac{(n-1)^2}{(n-2)(n-3)}. \quad (7)$$

2.3. The Experimental Conditions

All experiments were performed with the parallel plate geometry (Linkam CSS450, Linkam Scientific Instruments, Tadworth, UK), schematically shown in Figure 1. It consisted of two parallel quartz plates with a diameter of 36 mm, each in contact with flat silver heaters on the outside, with an observation window located at 7.5 mm radial position. The motion of the lower disc imposed a shearing stress field on the emulsion. Images were captured on the vorticity–velocity plane through the 2.8 mm observation window. The effects of the shearing and duration of the imposed shear flow were studied using a constant rotation rate with a speed control better than 1% of the rotational speed and using a sequence of microstructure measurements at spaced events in time. In the present work, a gap spacing of 0.1 mm between disks and a temperature 30 °C was used.

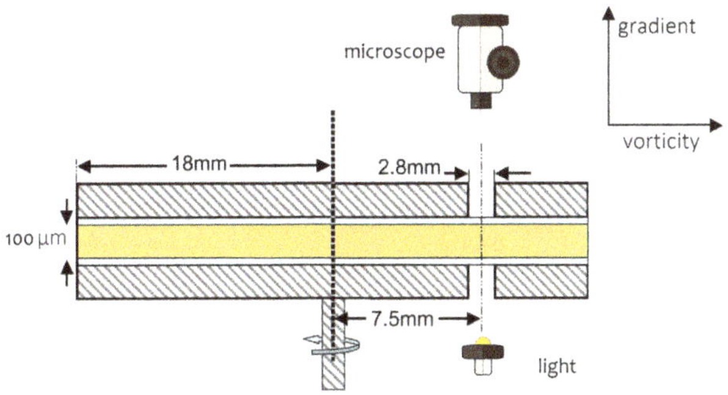

Figure 1. Schematic description of the shearing device: parallel-plate geometry with a diameter of 36 mm and a gap of 100 µm. The observation plane is described by the velocity–vorticity axes. The motion of the lower disc imposes a simple shearing stress field on the sample.

Emulsion structures were visualized using an optical microscope Nikon SMZ-U (Nikon Corp., Tokyo, Japan). Capture of images was carried out with a Nikon Digital Sight DS-2 mV camera in a bright field illumination arrangement. All images were processed with the ImageJ® software (U.S. National Institutes of Health, Bethesda, MD, USA), manually and automatically. Pre-processing of drop images included assigning a threshold value, converting the image to a binary map, followed by an automatic count of drops with circularity better than >0.94. The statistical methods are later applied after verifying the reliability and veracity of the data obtained by the image processing. Prior to the selection of an experimental run, the image capturing process was optimized in order to reduce the emulsion turbidity and proper exposure to light [33]. Thus, a full view image with the focus plane centered on the plane of the flow field was assured. In order to visualize the evolution of the microstructure, multiple images were taken towards the end of the shearing period and after the flow stopped and fully relaxed; images were spaced at intervals of 1 s for statistical analysis.

In order to obtain the droplet size distributions, mainly three operations were carried out (i.e., the image acquisition, pre-processing of digital images (involving cleansing, defining drop contours, etc.), and the statistical analysis of at least two (or multiple) images) to generate the frequency histograms. The observed droplets are those present only on the focal plane normal to the direction of the velocity-gradient. The depth of field for the optical arrangement is approximately 100 µm, about the size of the gap between plates; thus, most drops in the flow field were observed, while the width of the field view of the microscope on the focal plane was 2–3 times larger than the captured image, thus minimizing magnification ratio variations away from the central axis.

For $\tau < 0$—that is, using a gap of 100 µm during pre-shearing—images would have shown many small drops, with several of them frequently overlapping along the optical path, making the correct determination of the size or the count numbers of the drops almost impossible. The highest possible concentration (of small droplets) occurred at the onset of the flow ($\tau \equiv 0$). At this instant, the average size of the droplets implied a tight closeness of particles, and in order to be able to distinguish every individual droplet, data was only taken after $\sim$ 380 s. That is, even for the most concentrated emulsions, the image processing algorithm must assign a given size to each drop. It is only for $\dot{\gamma} = 0.75\ \mathrm{s}^{-1}$ that drops less than 5 µm are still observable at the bottom of the image (Figure 2a'). Understanding how this fraction of the population evolves (that is, attempting to elucidate the proper mechanism mediating this capture process) will require experiments with a smaller gap size.

For each subsequent constant shear rate flow section, the duration of the flow for ~400 s was sufficient to ensure the correct evaluation of the properties of the individual droplets at ~380 s after start-up of the steady flow. But as a shear rate of 1.5 s^{-1} was reached, only drops greater than 5 µm were observed for $\dot{\gamma} = 2.25$, 3.0, 3.75 and 4.5 s^{-1} (see Figure 2b'–f'). Under the prescribed conditions, all experiments addressed the changes in the size distributions of the particles as a function of (1) the shear rate; (2) the duration of the applied flow; and (3) the possible spatial variation of the distribution of particles parameters.

3. Results

Figure 2 shows the images captured during flow and immediately after flow (unprimed and primed labels, respectively—left most columns). The applied shear rates were from 0.75 to 4.5 s^{-1}, top to bottom, respectively, with the corresponding density function of drop sizes—right most column. The objective is to show several of the possible structures in an emulsion induced by flow and observed during these experiments.

Figure 2a shows many small drops, of diameter <5 µm and below the lower limit of the resolution threshold [23], for this optical arrangement; these drops are not included in the distribution for the shear rate of $\dot{\gamma} = 0.75\ \mathrm{s}^{-1}$. Indeed, these drops quickly disappear with increasing ($\tau \geq 400$), as observed in the bottom of the images of Figure 2b"–f". Possible experiments with a weaker shear rate, $\dot{\gamma} \leq 0.75\ \mathrm{s}^{-1}$, which may be most relevant for the smallest of drops, are not included here and are not considered to be relevant to the observed structural phenomena, which is the main objective of this work. Drops with a diameter larger than 45 µm are only a few and were not considered significant to the values of the parameters of the distribution. The relative frequency of drops larger than 45 µm is of the order of 0.00172.

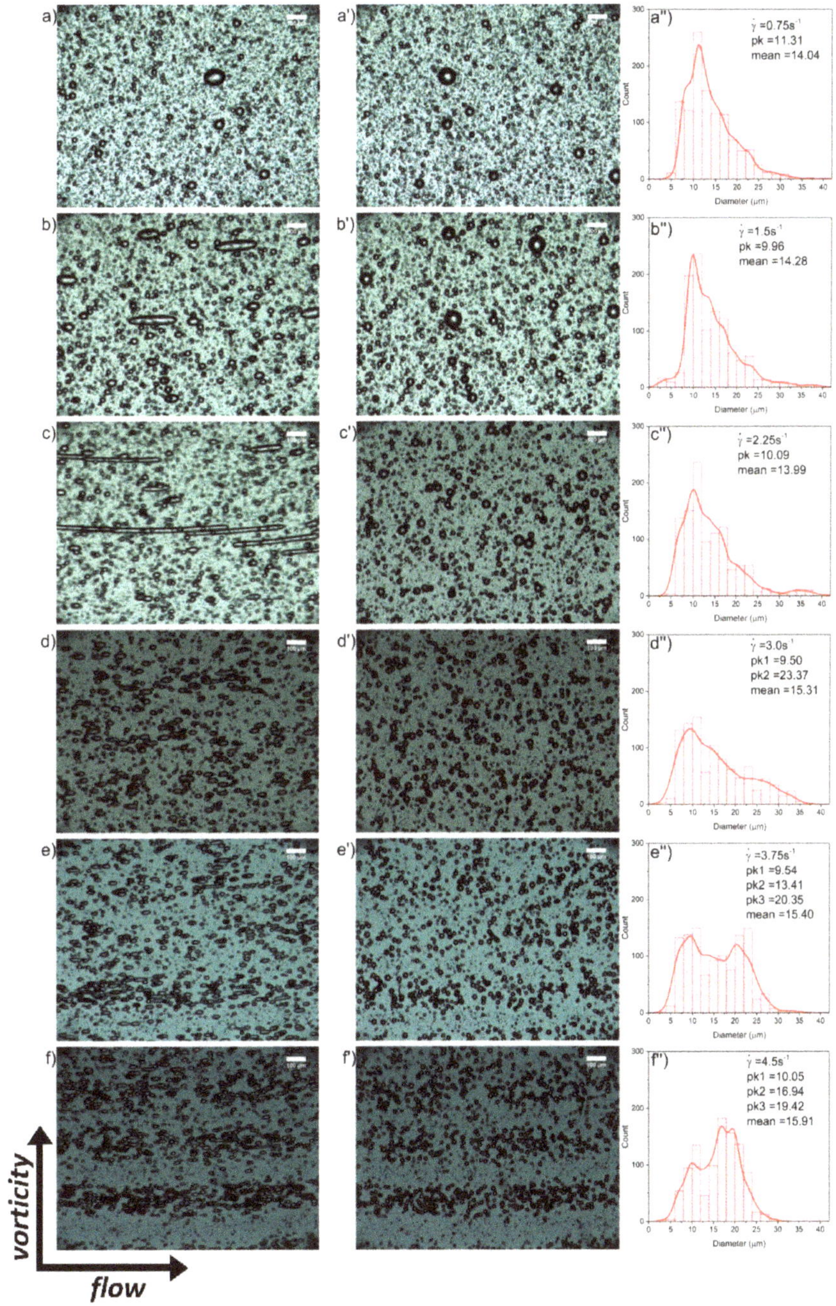

Figure 2. The sequence of images shows the effects due to increase of shear rates (top to bottom) on the evolution of the morphology—during flow, left, unprimed letters (**a**–**f**); after cessation of flow (**a'**–**f'**), center, and primed letters—for a (50/50) emulsion. For histograms—right plots, double primes (**a"**–**f"**)—pk*n* stands for the first, second, etc., peak of the distribution; mean is the average size of drops. All images are captured with the same magnification and at a temperature of 30 °C.

One of the first effects of the shearing that was observed that can readily be evaluated occurred for the smallest bin of the population, when the frequency of the smallest of drops (about 5–7.5 μm) decreased and the frequency of drops with about twice the volume (8–10 μm) clearly grew. This condition indicates that for flows with $\dot{\gamma} \leq 2.25\ s^{-1}$ the critical time for *coalescence of the smaller* drops already occurred. That is, at $\dot{\gamma} = 2.25\ s^{-1}$, the number of small drops is reduced within a few minutes, with pairs of drops generating larger drops of about twice the volume.

As shown in Figure 2f,f′, at larger shear rates, a poly-disperse emulsion with a multimodal distribution was generated. Here, Figure 2 shows that the average size of drops increased in time as the shear rate increased, accompanied by the formation of a (secondary) peak of larger drops and resulting in a *bi-modal* distribution. The same coalescence phenomenon appears to dominate the onset of the bi-modal distribution shown in Figure 2f″ from smaller size drops. At this flow rate—$\dot{\gamma} = 4.5\ s^{-1}$—the collision of drops with an average diameter close to 12–14 μm gave rise to drops of about 16.5 μm (double the volume of the smaller drops). From the histogram in Figure 2f″, it is clear that a new population phenomenon was at play: both frequencies of the 10–12 μm and 12–14 μm bins decreased significantly and simultaneously. It appears that pairs of drops of size 10–14 μm may coalesce rapidly, until a more stable drop size is attained, both bins contributing to the appearance of the second and third peaks in the distribution at about 14–20 μm.

Interestingly, Figure 2f also shows (a) the onset of *pearl collar* structures—several drops of similar diameter, evenly spaced and roughly aligned along the flow direction—with ellipsoid-like drop shapes and a waist close to ~15 μm; and (b) a banded, structured emulsion along the flow direction and perpendicular to the vorticity axis. The surprising feature of the collars and bands were their persistent lengths, several times the characteristic length scale of previous phenomena and several times the thickness of the channel (about 300 μm).

In these experiments, a few large drops were also observed—larger than 40 μm in diameter, with an ellipsoid-like shape with an averaged waist size ~35 μm—at the onset of the flow regime that did not show up initially in the upper tail of the distribution. In Figure 2c, the highly elongated drops—similar to wire-structures—could be attributed to possibly larger drops that had reached a very large deformation, only possibly due to confinement effects by neighboring drops and by the presence of the cell walls, simultaneously.

Table 1 presents the data used for each experiment of the most relevant parameters characterizing these distributions. Thus, approximately 1000 drops were captured for each drop size distribution, and several *measures of the diameter* were calculated. Of particular relevance are the mean ($D_{1,0}$), as well as the Number-area and Number-volume mean diameter. The relevance of these numbers is further expanded upon in the Section 4.

Table 1. Drop size distribution statistics for the histograms shown in Figure 2.

Shear Rate (s^{-1})	Total Number Drops	Diameter Mean (μm)	Diameter Median (μm)	$D_{4,3}$ (μm)	$D_{3,2}$ (μm)	$D_{2,0}$ (μm)	$D_{1,0}$ (μm)	Distribution Properties	
								Skewness	Kurtosis
0.75	1148	14	12.9	21.1	18.6	15	13.9	1.2	1.82
1.5	1078	14.3	13.2	22.8	19.7	15.6	14.4	1.32	2.46
2.25	1084	14	13.2	24.3	20.6	15.6	14.2	1.44	2.49
3	1015	15.3	14.1	24.7	22.2	17.2	15.5	0.73	−0.39
3.75	1135	15.4	15	21.9	20.3	16.8	15.6	0.26	−0.86
4.5	1034	15.9	16.4	20	19	16.9	16.2	−0.12	−0.61

$D_{4,3}$ is the volume or mass moment mean also known as the De Broucker mean diameter. $D_{3,2}$ is the surface area moment mean or the Sauter mean diameter (SMD). $D_{2,0}$ is the number-area mean diameter. Kurtosis measures the relative weight of the tails with respect to the central portion of the distribution.

In Figure 3, the drop size distributions are shown with their corresponding smoothed kernels for all shear rates studied in this work. The horizontal axis is the bin average diameter and the vertical axis is the drop count. It is observed in Figure 3a–c that these size distributions are characterized by a single peak and positive Kurtosis—with a single peak and relatively high weight of the distribution tails (see Table 1). Please note that for a flow cell with a gap of 100 µm, drops with diameters as large as 100 µm should be observed; although, very few are seen during these experiments, which do not significantly alter the Kurtosis of the distribution. Most drop diameters are constrained to less than 40 µm. In Figure 3d, the maximum value of $d_V = D_{4,3}$ is reached (blue vertical bar). In Figure 3e,f, a rapid decrease in the value of $D_{4,3}$ (Figure 4) is shown. This effect is mainly due to a rather broad peak, which here becomes evident by the presence of a second peak that emerges from right to left, and a Kurtosis measure that changes sign. For the drop size distribution generated when $\dot{\gamma} = 4.5\ \mathrm{s}^{-1}$, the skewness parameter becomes negative indicating a larger (secondary) peak at higher diameters.

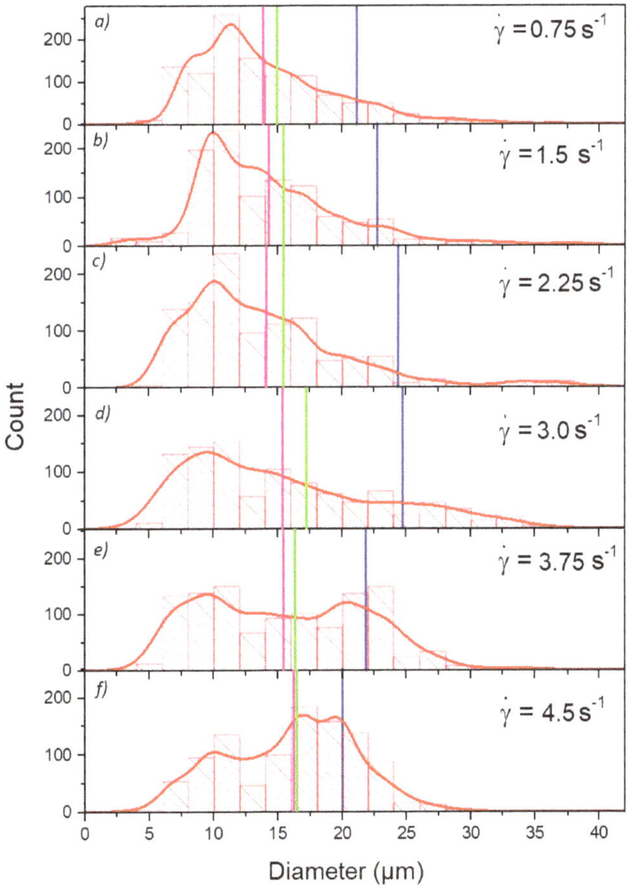

Figure 3. Droplet size distribution and the smoothing kernels for all shear rates studied. For the first three (lower) shear rates (**a–c**), the size distributions are characterized by a single peak and positive Kurtosis [34–37]; with rather broad distributions. The magenta bar corresponds to the mean diameter for the distribution, the green bar to the $D_{2,0}$ value, and the blue bar corresponds to the value of $D_{4,3}$. In histogram (**d**), the maximum value of $D_{4,3}$ is reached (see Table 1). Histograms (**e**,**f**) show a rapid decrease in the value of $D_{4,3}$, characteristic of distributions with a high top and a smaller contribution from outliers.

Figure 4 presents the evolution of the mean drop size using the suggested averages (values shown in Table 1). Figure 4a presents the diameter evolution with respect to the dimensionless time (or as a function of the applied shear *strain* by the flow); Figure 4b corresponds to the same data vs. the applied shear rate. It is clear that $D_{1,0}$ and $D_{2,0}$ remain essentially constant at 14 and 15.5 μm, respectively, for times less than $\tau \leq 800$, abruptly increasing to means of 15.5 and 17 μm, respectively, under a faster flow and a longer duration of the flow.

Figure 4c,d correspond to the normalized average size ($D_{4,3}/D_{1,0}$ and $D_{2,0}/D_{1,0}$) versus the dimensionless time and the *accumulated dimensionless time* (total imposed strain). Therefore, the base line for these two plots corresponds to $D_{1,0}$, a two-step function with rather constant lower and upper values (i.e., 14 μm and 15.5 μm (see Figure 4a)). The droplets generated during the preparation of the emulsion grew constantly within the initial 675 units of dimensionless time and then evolved rapidly to an average mean diameter of 15.5 μm. In most experiments, close to 300–10,000 units of dimensionless time are necessary to reach a reproducible size distribution of the emulsions. The longer times are especially the case for global anisotropic structures, which may require a *total dimensionless time* longer than 8000, as shown in Figure 4d.

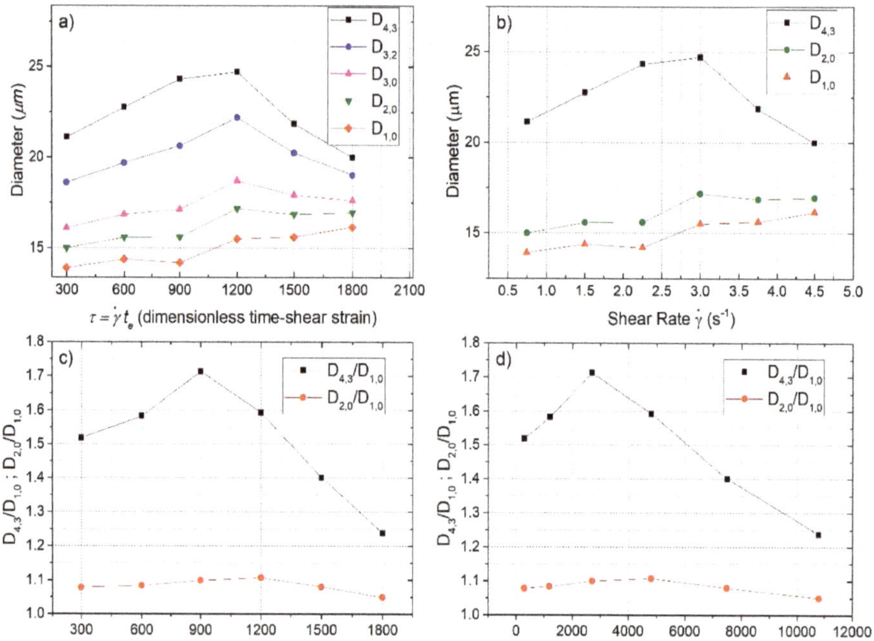

Figure 4. (**a**) Mean diameters of the distributions—$D_{1,0}$ up to $D_{4,3}$—vs. the dimensionless time τ_i incurred during each flow regime; (**b**) $D_{1,0}$, $D_{2,0}$, and $D_{4,3}$ vs. the applied shear rate during each regime; (**c**,**d**) are, respectively, the dimensionless *normalized* mean drop size vs dimensionless time in each cycle and for the *total duration of the flow* $\sum \dot{\gamma}_i t_i$.

4. Discussion

The distributions of drops appear to change mainly due to two processes, coalescence and rupture, with little evidence of Oswald ripening. The lack of evidence for the latter may be due to the fact that drops larger than 35–40 μm were not observed. In contrast, the presence of the coalescence phenomenon can be inferred at several instances in the evolution of the distributions, which may be due to different hydrodynamic processes albeit operating simultaneously. As pointed out in the description of the coalescence kinetics, the most likely case corresponds to two of the smaller drops (about equal size) generating a drop with about twice their volume. This was clearly the case for the smaller drops in Figure 3a–d. But coalescence may also be the dominant mechanism for the reduction of probability observed in the first peak and in the dawning of the second peak. During collision of these larger drops, those of $\sim$ 10 μm disappear, with the 15 μm bin growing and during a subsequent coalescence, becoming drops of about 20 μm. Coalescence seems to be the dominant mechanism for critical dimensionless time less than $\tau \leq$ 700, with an associated critical shear rate of about $\dot{\gamma} \leq$ 2.25 s^{-1}. Coalescence occurs simultaneously with rupture up to $\tau \approx 8000 \dot{\gamma} \leq$ 4.5 s^{-1}.

A second possible mechanism for the onset of the secondary peak is break up, particularly for the heavier tail, with the disappearance of the much larger drops (less than 40 μm) producing smaller drops of about half their volume (i.e., $\sim$27 μm (as shown when $\dot{\gamma}$ = 3.0–3.75 s^{-1} (Figure 3d,f))). These drops, in turn, break up again, producing an increase in frequency for drops sizes of about 22 μm. The rupture mechanism appears to depend on dimensionless time and on a critical shear rate of the order of $\dot{\gamma}$ = 3.0–3.75 s^{-1}, not being observed for smaller drops. The simultaneous contribution of these two mechanisms makes it possible to explain the appearance of the second peak of the bimodal distribution.

Figure 4 presents the evolution of the mean drop size (with five different measures) versus its shear rate history. Initially, droplets generated during the preparation of the emulsion grew constantly, from sizes less than the threshold of the image processing, but attained values that can be readily evaluated with the statistical measures proposed for $\tau \geq$ 300, as shown in Figure 4. The mean value $D_{1,0}$ remained at a constant value for shear rates less than $\dot{\gamma} \leq$ 2.25 s^{-1}—the same behavior is observed for $D_{2,0}$—while $D_{3,2}$ $D_{4,3}$ increased linearly with dimensionless time up to 900τ units, followed by a monotonic decrease of the mean size for the total accumulated dimensionless time $\tau \geq$ 3000.

For dimensionless time between 675–900 and under an increment of $\dot{\gamma}$ > 2.25 s^{-1}, the mean value $D_{1,0}$ (and $D_{2,0}$) appeared to remain at a new constant value of $D_{1,0}$ = 15.5 μm (and $D_{2,0}$ = 17.5 μm) approximately. This jump in mean diameter correlated with the time and shear rate in these experiments. An interesting observation is that the ratio of these two means was essentially constant for the complete range of values of the shear rate (see Figure 4c) regardless of the dimensionless time or shear rate. In contrast, measures $D_{3,2}$ and $D_{4,3}$ did change significantly (−25%) after $\tau \geq$ 675 or under an increase of $\dot{\gamma}$ > 3.0 s^{-1}. This abrupt collapse of $D_{3,2}$ and $D_{4,3}$ values correlated with the behavior of other statistical measures, indicating a dominant breakup mechanism of the larger drops. The rupture mechanism was most clearly associated with the appearance of pairs of satellite drops (of about half volume) that did not appear to influence the low end of the distribution. Hence, under these flows, the drops attained deformations associated with dumbbell shapes before breaking up, with a few of them reaching a significantly elongated waist capable of generating smaller satellite drops.

In most cases, 100–10,000 units of strain are necessary to reach a size distribution of the emulsion with low tail populations. Figure 4d shows a *total time* longer than $\tau \geq$ 8000, a time scale that is representative of the anisotropic global structures observed in this emulsion. By increasing the shear rate, the arithmetic average value increased at a rather constant rate. However, the position of the peaks was not affected in a forceful way and appeared to be independent of time, contrary to the behavior of the width of each of the main peaks, which became broader over time while keeping the area under the curve approximately equal. This phenomenon may have been the result of the disappearance of the smallest droplets, as shown in Figure 3f. For the large drops, the formation of a secondary peak from right to left was observed (see Figure 3d–f), resulting in the decrease of the height of the first peak and the increase of its width and vice versa for the second peak.

Figure 4d shows a *total time* longer than $\tau \geq$ 8000. In other to evaluate in detail the *global structure* shown in Figure 2e,f, associated with τ $\sim$ 8000 global times, future experiments may require times significantly longer than the length of time considered here, as well as a new statistical analysis for evaluation of the anisotropic character of the distribution of particles in the sample.

5. Conclusions

This emulsion showed good stability when left to rest. However, under flow conditions, the size distribution changed significantly in terms of the mean values, as well as other characteristics, such as mono- to bimodal distributions or measures of skewness and kurtosis. Besides coalescence and rupture as possible explanations of the observed behavior at lower shear rates, other possible structures are readily observable, such as necklaces and banded structures. Evidence of necklaces appear at a transition $\dot{\gamma} \sim 3.0\,\text{s}^{-1}$ in a single band to transient chains of drops in the flow direction and in several layers at a shear rate of $\dot{\gamma} \sim 3.75\,\text{s}^{-1}$.

These experiments also elucidate two possible mechanisms for the evolution of the size distribution of drops in highly concentrated emulsions at low shear rates. The first is coalescence, dominant for smaller drops at low shear rates and short times. The arithmetic mean of 14.0–14.3 μm under a shear rate of about 0.75–1.5 s^{-1} can be explained by coalescence of the smallest drops, while indicating a poor stability to flow induced perturbations. The behavior of the small drops indicates a high frequency of close neighbor encounters, promoting coalescence during this stage. Due to the fact that the efficiency of coalescence decreases with the increasing difference in the size of the colliding drops [2,24], coalescence tends to be less likely at longer times. These results are in agreement with the qualitative works of Grizzuti and Bifulco [25] and Rusu and Peuvrel-Disdier [23].

The weak increase of the $D_{1,0}$ value of the distribution as the shear rate grows may indicate that it is independent of time, which is in contrast to the behavior of the width of each of the main peaks, which increase as time passes while keeping the area under the curve approximately constant. Observing the evolution of the peaks and width of peaks in Figure 3d–f, it becomes evident that the largest drops rupture, giving rise to smaller ones. These dynamics may induce the appearance of the second peak (of larger drops) in Figure 3f. Here, Figure 4c,d show that the $D_{4,3}$ measure decreases significantly relative to the mean value when the tail of the distribution for large drops is decreasing and rupture is a prevalent mechanism. This hypothesis may be correct; although, we have not addressed this point in detail, we may do it in future work. The tails of the distribution are less dominant as shown by the skewness and kurtosis values.

Author Contributions: J.M.L. and E.G. conceived of and designed the experiments; J.M.L. performed the experiments; J.M.L. and E.G. analyzed the data; E.G. contributed reagents/materials/analysis tools; and J.M.L. and E.G. wrote the paper.

Funding: This research was funded by DGAPA-UNAM, Mexico; grant number PAPIIT-IN114618.

Acknowledgments: We thank the reviewers for their insightful comments and suggestions. J.M.L. thanks for financial support throughout his Ph.D. studies in Materials Science and Engineering, UNAM, to Consejo Nacional de Ciencia y Tecnología (CONACYT, Mexico).

Conflicts of Interest: The authors declare no conflict of interest. Sponsors had no role in the design of the study; in the data collection, analyses, or interpretation of data; in the writing of the manuscript; or in the decision to publish the results.

References

1. Berg, J.C. Fluid interfaces and capillarity. In *An Introduction to Interfaces & Colloids: The Bridge to Nanoscience*; World Scientific: Singapore, 2010; pp. 30–80, ISBN 978-9-81-429307-5.
2. Park, J.Y.; Blair, L.M. The effect of coalescence on drop size distribution in agitated liquid-liquid dispersion. *Chem. Eng. Sci.* **1975**, *30*, 1057–1064. [CrossRef]
3. Sundararaj, U.; Macosko, C.W. Drop breakup and coalescence in polymer blends: The effects of concentration and compatibilization. *Macromolecules* **1995**, *28*, 2647–2657. [CrossRef]
4. Leal, L.G. Flow induced coalescence of drops in a viscous fluid. *Phys. Fluids* **2004**, *16*, 1833–1851. [CrossRef]
5. De Bruijn, R.A. Tipstreaming of drops in simple shear flows. *Chem. Eng. Sci.* **1993**, *48*, 277–284. [CrossRef]
6. Rumscheidt, F.D.; Mason, S.G. Particle motions in sheared suspensions XII. Deformation and burst of fluid drops in shear and hyperbolic flow. *J. Colloid Sci.* **1961**, *16*, 238–261. [CrossRef]

7. Rumscheidt, F.D.; Mason, S.G. Particle motions in sheared suspensions XI. Internal circulation in fluid droplets (experimental). *J. Colloid Sci.* **1961**, *16*, 210–237. [CrossRef]
8. Delaby, I.; Ernst, B.; Germain, Y.; Muller, R. Droplet deformation in polymer blends during uniaxial elongational flow: Influence of viscosity ratio for large capillary numbers. *J. Rheol.* **1994**, *38*, 1705–1720. [CrossRef]
9. Caserta, S.; Sabetta, L.; Simeone, M.; Guido, S. Shear-induced coalescence in aqueous biopolymer mixtures. *Chem. Eng. Sci.* **2005**, *60*, 1019–1027. [CrossRef]
10. Frith, W.J.; Lips, A. The rheology of concentrated suspensions of deformable particles. *Adv. Colloid Interface Sci.* **1995**, *61*, 161–189. [CrossRef]
11. Vinckier, I.; Moldenaers, P.; Terracciano, A.M.; Grizzuti, N. Droplet size evolution during coalescence in semiconcentrated model blends. *AIChE J.* **1998**, *44*, 951–958. [CrossRef]
12. Lyu, S.P.; Bates, F.S.; Macosko, C.W. Coalescence in polymer blends during shearing. *AIChE J.* **2000**, *46*, 229–238. [CrossRef]
13. Taylor, G.I. The viscosity of a fluid containing small drops of another fluid. *Proc. R. Soc. Lond. A.* **1932**, *138*, 41–48. [CrossRef]
14. Utracki, L.A.; Shi, Z.H. Development of polymer blend morphology during compounding in a twin-screw extruder. Part I, Droplet dispersion and coalescence—A review. *Polym. Eng. Sci.* **1992**, *32*, 1824–1833. [CrossRef]
15. Torza, S.; Cox, R.G.; Mason, S.G. Particle motions in sheared suspensions XXVII. Transient and steady deformation and burst of liquid drops. *J. Colloid Interface Sci.* **1972**, *38*, 395–411. [CrossRef]
16. Grace, H.P. Dispersion phenomena in high viscosity immiscible fluid systems and application of static mixers as dispersion devices in such systems. *Chem. Eng. Commun.* **1982**, *14*, 225–277. [CrossRef]
17. De Bruijn, R.A. Deformation and Breakup of Drops in Simple Shear Flows. Ph.D. Thesis, Eindhoven University of Technology, Eindhoven, The Netherlands, 1989.
18. Chesters, A. Modelling of coalescence processes in fluid-liquid dispersions: A review of current understanding. *Chem. Eng. Res. Des.* **1991**, *69*, 259–270.
19. Abid, S.; Chesters, A.K. The drainage and rupture of partially-mobile films between colliding drops at constant approach velocity. *Int. J. Multiph. Flow* **1994**, *20*, 613–629. [CrossRef]
20. Coulaloglou, C.A.; Tavlarides, L.L. Description of interaction processes in agitated liquid-liquid dispersions. *Chem. Eng. Sci.* **1977**, *32*, 1289–1297. [CrossRef]
21. Ramic, A.J.; Stehlin, J.C.; Hudson, S.D.; Jamieson, A.M.; Manas-Zloczower, I. Influence of block copolymer on droplet breakup and coalescence in model immiscible polymer blends. *Macromolecules* **2000**, *33*, 371–374. [CrossRef]
22. Liao, Y.; Lucas, D. A Literature review on mechanisms and models for the coalescence process of fluid particles. *Chem. Eng. Sci.* **2010**, *65*, 2851–2864. [CrossRef]
23. Rusu, D.; Peuvrel-Disdier, E. In Situ characterization by small angle light scattering of the shear-induced coalescence mechanisms in immiscible polymer blends. *J. Rheol.* **1999**, *43*, 1391–1409. [CrossRef]
24. Lyu, S.; Bates, F.S.; Macosko, C.W. Modeling of coalescence in polymer blends. *AIChE J.* **2002**, *48*, 7–14. [CrossRef]
25. Grizzuti, N.; Bifulco, O. Effects of coalescence and breakup on the steady-state morphology of an immiscible polymer blend in shear flow. *Rheol. Acta* **1997**, *36*, 406–415. [CrossRef]
26. Minale, M.; Mewis, J.; Moldenaers, P. Study of the morphological hysteresis in immiscible polymer blends. *AIChE J.* **1998**, *44*, 943–950. [CrossRef]
27. Minale, M.; Moldenaers, P.; Mewis, J. Effect of shear history on the morphology of immiscible polymer blends. *Macromolecules* **1997**, *30*, 5470–5475. [CrossRef]
28. Bentley, B.J.; Leal, L.G. An experimental investigation of drop deformation and breakup in steady, two-dimensional linear flows. *J. Fluid Mech.* **1986**, *167*, 241–283. [CrossRef]
29. Jansen, K.M.; Agterof, W.G.; Mellema, J. Droplet breakup in concentrated emulsions. *J. Rheol.* **2001**, *45*, 227–236. [CrossRef]
30. Guido, S.; Villone, M. Measurement of interfacial tension by drop retraction analysis. *J. Colloid Interface Sci.* **1999**, *209*, 247–250. [CrossRef] [PubMed]
31. Silverman, B.W. The kernel method for univariate data. In *Density Estimation for Statistics and Data Analysis*, 1st ed.; Routledge: New York, NY, USA, 1986; pp. 34–74, ISBN 0412246201.

32. Scott, D.W. Kernel Density Estimators. In *Multivariate Density Estimation: Theory, Practice, and Visualization*, 2nd ed.; John Wiley & Sons: Hoboken, NJ, USA, 2015; pp. 137–216, ISBN 1118575482.
33. Caserta, S.; Simeone, M.; Guido, S. 3D optical sectioning and image analysis of particles in biphasic systems. *Microsc. Anal.* **2005**, *19*, 9–11.
34. Wang, H. Path factors of bipartite graphs. *J. Graph Theory* **1994**, *18*, 161–167. [CrossRef]
35. Kurtosis in Probability Distributions. Available online: https://en.wikipedia.org/wiki/Kurtosis (accessed on 22 April 2018).
36. Kim, T.H.; White, H. On more robust estimation of skewness and kurtosis. *Financ. Res. Lett.* **2004**, *1*, 56–73. [CrossRef]
37. Moments. Available online: https://www.originlab.com/doc/X-Function/ref/moments (accessed on 22 April 2018).

fluids MDPI

Article

Direct Numerical Simulation of Particles in Spatially Varying Electric Fields †

Edison Amah [1], Muhammad Janjua [2] and Pushpendra Singh [1,*]

1 Department of Mechanical Engineering, New Jersey Institute of Technology, Newark, NJ 07102, USA; eca8@njit.edu
2 Department of Mechatronics and Mechanical Engineering, Higher Colleges of Technology, P. O. Box 15825, Dubai, UAE; janjua@gmail.com
* Correspondence: singhp@njit.edu; Tel.: +1-973-596-3326

† This paper is a modified version of our paper "Direct Numerical Simulations (DNS) of Particles in Spatially Varying Electric Fields" published at the ASME 2014 4th Joint US-European Fluids Engineering Division Summer Meeting collocated with the ASME 2014, Chicago, IL, USA, 3–7 August 2014.

Received: 15 June 2018; Accepted: 17 July 2018; Published: 24 July 2018

Abstract: A numerical scheme is developed to simulate the motion of dielectric particles in the uniform and nonuniform electric fields of microfluidic devices. The motion of particles is simulated using a distributed Lagrange multiplier method (DLM) and the electric force acting on the particles is calculated by integrating the Maxwell stress tensor (MST) over the particle surfaces. One of the key features of the DLM method used is that the fluid-particle system is treated implicitly by using a combined weak formulation, where the forces and moments between the particles and fluid cancel, as they are internal to the combined system. The MST is obtained from the electric potential, which, in turn, is obtained by solving the electrostatic problem. In our numerical scheme, the domain is discretized using a finite element scheme and the Marchuk-Yanenko operator-splitting technique is used to decouple the difficulties associated with the incompressibility constraint, the nonlinear convection term, the rigid-body motion constraint and the electric force term. The numerical code is used to study the motion of particles in a dielectrophoretic cage which can be used to trap and hold particles at its center. If the particles moves away from the center of the cage, a resorting force acts on them towards the center. The MST results show that the ratio of the particle-particle interaction and dielectrophoretic forces decreases with increasing particle size. Therefore, larger particles move primarily under the action of the dielectrophoretic (DEP) force, especially in the high electric field gradient regions. Consequently, when the spacing between the electrodes is comparable to the particle size, instead of collecting on the same electrode by forming chains, they collect at different electrodes.

Keywords: dielectrophoresis; direct numerical simulations; Maxwell stress tensor method; point-dipole method; distributed Lagrange multiplier method

1. Introduction

In recent years, considerable attention has been given to understanding the behavior of particles suspended within liquids because of their importance in a wide range of applications, e.g., self-assembly of micron to nano-structured materials, separation and trapping of biological particles [1], stabilization of emulsions, and the formation of fluids with adjustable rheological properties, etc. [2–11]. Future progress in these and related areas will critically depend upon our ability to accurately control the arrangement of particles for a range of particle types and sizes, including those of uncharged particles.

When a dielectric particle is subjected to a spatially non-uniform electric field it experiences an electrostatic force, called the dielectrophoretic (DEP) force. The DEP force arises because the particle becomes polarized and the polarized particle (or dipole) placed in a spatially varying electric field experiences a net force. This phenomenon itself is referred to as dielectrophoresis [2]. The force can be in the direction of the gradient of electric field or in the opposite direction. For a positively polarized particle the force is in the direction of the electric field gradient and for a negatively polarized particle the force is in the opposite direction to the gradient. In addition, the particle interacts electrostatically with other particles which can be modeled as dipole-dipole interactions. The dipole-dipole interactions are present even in a uniform electric field.

Dielectrophoresis is an important technique because it can be used to manipulate uncharged particles. Also, since the DEP force depends on the dielectric properties of the particle, the force is different on different kinds of particles. Therefore, in principle, DEP force can be used to separate particles with different dielectric properties [3]. For example, cancer cells can be separated from normal cells as they have different dielectric properties [12–15].

Here we present a direct numerical simulation (DNS) method based on the finite element method which can be used to study the motion of dielectric particles suspended in a dielectric liquid. The electric field can be uniform or nonuniform, e.g., the field in a dielectrophoretic cage is nonuniform. Direct numerical simulations are important for understanding the dependence of the DEP force-induced motion on the fluid and particle properties, and also for designing efficient microfluidic devices.

The point dipole (PD) approximation [2,14] which has been used in many past studies assumes that the particle is small compared to the length scale over which the electric field varies, and that the distance between the particles is much larger than the particle diameter. In this approximation, the electric force acting on a particle consists of two components, one due to dielectrophoresis, and the other due to the particle-particle interactions. The accuracy of the PD approximation decreases when the electric field varies significantly over a length scale comparable to the particle size and also when the distance between the particles is small.

The accuracy of the PD approximation can be improved by including higher order multipole moment terms in the force [16]. However, as shown in Reference [17], even when the multipole terms are included, it lacks the completeness of the Maxwell Stress Tensor (MST) method. In the latter method, the force is computed from the electric field that accounts for the presence of all particles in the computational domain and so the computed force incorporates multibody interactions [17]. The MST method is computationally demanding because the electric field has to be computed at each time step as the field changes when the particles move [18–20]. In Reference [21], an energy based method, which accounts for the presence of particles, was used to compute the force on the particles in a uniform electrical field. In this method, both far and near field interactions are included in the particle-particle force.

During the past 10 years, several new DNS approaches have been developed to model dielectrophoresis [22–29]. A scheme to study ac dielectrophoresis is described in Reference [30]. In this scheme, the ac electric field is obtained by solving the quasi-electrostatic form of the Maxwell equation from which the DEP force acting on the particles is obtained and the motion of particles in the fluid is simulated using the immersed boundary method. The scheme was used to study the motion of levitated particles and to evaluate the accuracy of the point dipole method. It was noted that in a given domain the accuracy of the PD method decreases with increasing particle size and that the accuracy diminishes in a high electric field gradient region. A numerical approach to study chain formation by particles with dissimilar electrical conductivities, sizes and shapes is described in [24]. An arbitrary Lagrangian-Eulerian (ALE)-based method was used in Reference [25,26] to study ac and dc dielectrophoresis in two-dimensions. The focus of this work was on understanding the mechanisms involved in particle assembly. A scheme based on the boundary-element method (BEM) was developed in Reference [27] to solve the coupled electric field, fluid flow and particle motion problems in the Stokes flow limit. A PD-based method in which the induced dipole moments within the particles are computed by accounting for the nearby particles, was developed in Reference [28].

There are several other studies in which only the DEP force lines in the devices were considered and the particle trajectories were estimated without considering the fluid-particle interactions [30,31].

Recently, in Reference [28], a multiphase model-based approach has been developed to study the motion of a large number of particles. The dielectrophoretic force in this approach is calculated using the point dipole approach, and the drag acting on the particles is modelled by Stokes law. The fluid flow induced due the motion of particles is not considered in this model. These model-based simulations allow us to study systems with thousands of particles which, at present, is not possible when a DNS approach is employed. However, a model-based approach may not accurately capture the underlying physics, and so the DNS or experimental results are needed to validate it.

In our DNS method, the exact equations governing the motion of the fluid and the particles are solved and subjected to the appropriate boundary conditions. The scheme uses a distributed Lagrange multiplier (DLM) method to enforce the rigid body motion constraint in the region occupied by the particles [32] and the electric force acting on the particles is computed using the point-dipole and Maxwell stress tensor approaches. One of the focuses of this paper is on the case where the particle size is comparable to the length scale over which the electric field varies. It is shown that the tendency of particles to form chains diminishes with increasing particle size, as the DEP force increases faster than the particle-particle interaction force. Thus, although particles come together to form chains when they are away from the electrodes, they may not collect at the same electrode as the neighboring electrode pulls some particles away from the first electrode. This happens because the presence of particles modifies the electric field, which changes the magnitude and direction of the DEP force.

2. Methods

The motion of particles suspended in fluids under the action of electric forces is governed by highly nonlinear coupled partial differential equations. We must solve the Navier-Stokes equation for the fluids which are coupled with the equation of motion for the particles and satisfy the boundary conditions. In addition, we must determine the electric field and calculate the electric force by integrating the Maxwell stress tensor over the surface of particles. The flow field and the electric field must be resolved at a scale smaller than the size of particles as well as in the gap between the particles. Let us assume that there are N solid particles in the domain denoted by Ω and denote the interior of the i th particle by $P_i(t)$, and the domain boundary by Γ (see Figure 1). The momentum and mass conservation equations are

$$\rho_L\left(\frac{\partial \mathbf{u}}{\partial t} + \mathbf{u}\cdot\nabla\mathbf{u}\right) = -\nabla p + \nabla\cdot(2\eta\mathbf{D}) \text{ in } \Omega\backslash\overline{P(t)} \tag{1}$$

$$\nabla\cdot\mathbf{u} = 0 \text{ in } \Omega\backslash\overline{P(t)} \tag{2}$$

where $\mathbf{u}$ is the fluid velocity, p is the pressure, η is the dynamic viscosity of the fluid, ρ_L is the fluid density, $\mathbf{D}$ is the symmetric part of the velocity gradient tensor. The boundary conditions on the domain and the particle boundaries are:

$$\begin{cases} \mathbf{u} = \mathbf{u}_L \; on \; \Gamma \\ \mathbf{u} = \mathbf{U}_i + \omega_i \times \mathbf{r}_i \; on \; \Gamma_{Pi}, \; i = 1,\ldots,N \end{cases} \tag{3}$$

where $\mathbf{U}_i$ and $\boldsymbol{\omega}_i$ are the linear and angular velocities of the i th particle, respectively, and $\Gamma_{Pi} = \partial P_i(t)$ is the boundary of the i th particle.

The above equations are solved using the initial condition $\mathbf{u}|_{t=0} = \mathbf{u}_0$, where $\mathbf{u}_0$ is the known initial value of the velocity.

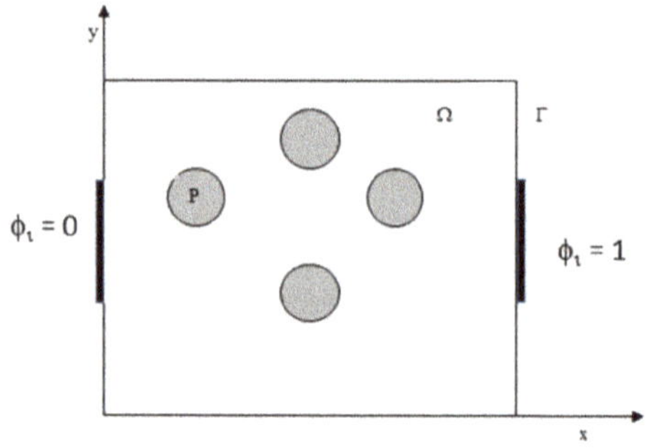

Figure 1. A typical rectangular prism shaped domain used in our simulations. The fluid velocity on the domain walls is assumed to be zero and the electrode's potential is specified.

The linear velocity $\mathbf{U}_i$ and the angular velocity ω_i of the i th particle are governed by

$$m_i \frac{d\mathbf{U}_i}{dt} = \mathbf{F}_i + \mathbf{F}_{E,i} \tag{4}$$

$$I_i \frac{d\omega_i}{dt} = \mathbf{T}_i + \mathbf{T}_{E,i} \tag{5}$$

$$\begin{cases} \mathbf{U}_i|_{t=0} = \mathbf{U}_{i,0} \\ \omega_i|_{t=0} = \omega_{i,0} \end{cases} \tag{6}$$

where m_i and I_i are the mass and moment of inertia of the i th particle, respectively. $\mathbf{F}_i$ and $\mathbf{T}_i$ are the hydrodynamic force and torque acting on the i th particle, respectively, $\mathbf{F}_{E,i}$ is the electrostatic force acting on the i th particle and $\mathbf{T}_{E,i}$ is the electrostatic torque acting on the i th particle. In this paper, our focus is on the case of spherical particles, and so we do not need to keep track of the particle orientation. The particle velocities are then used to update the particle positions

$$\frac{d\mathbf{X}_i}{dt} = \mathbf{U}_i \tag{7}$$

$$\mathbf{X}_i|_{t=0} = \mathbf{X}_{i,0} \tag{8}$$

where $\mathbf{X}_{i,0}$ is the position of the i th particle at time $t = 0$.

2.1. Formulation of Algorithms

2.1.1. Point-Dipole and Maxwell Stress Tensor Approaches

In the PD approximation, the electric force acting on a particle is obtained by computing the net particle-particle interaction force and adding it to the dielectrophoretic force. The time averaged dielectrophoretic force in an AC electric field is [2,14]

$$\mathbf{F}_{DEP} = 2\pi a^3 \varepsilon_0 \varepsilon_c \beta \, \nabla \mathbf{E}^2 \tag{9}$$

where a is the particle radius, ε_c is the dielectric constant of the fluid, $\varepsilon_0 = 8.8542 \times 10^{-12}$ F/m is the permittivity of free space and $\mathbf{E}$ is the RMS (root mean square) value of the electric field. Equation (9) is also valid for a DC electric field, in which case **E** is simply the electric field intensity. The coefficient $\beta(\omega)$ is the real part of the frequency-dependent Clausius-Mossotti factor, given by

$$\beta(\omega) = \mathrm{Re}\left(\frac{\varepsilon_p^* - \varepsilon_c^*}{\varepsilon_p^* + 2\varepsilon_c^*}\right) \tag{10}$$

where ε_p^* and ε_c^* are the frequency dependent complex permittivities of the particle and fluid, respectively. Here $\varepsilon^* = \varepsilon - j\sigma/\omega$ is the complex permittivity, where ε and σ are the permittivity and conductivity, and $j = \sqrt{-1}$. Notice that the value of Clausius-Mossotti factor is between -0.5 and 1.0.

In a non-uniform electric field, the particle-particle interaction force on the i th particle due to the j th particle is obtained by accounting for the spatial variation of the electric field [33–36]

$$\mathbf{F}_D = \frac{12\pi\varepsilon_0\varepsilon_c\beta^2 a^6}{r^4}\left(\mathbf{r}_{ij}(\mathbf{E}_i\cdot\mathbf{E}_j) + (\mathbf{r}_{ij}\cdot\mathbf{E}_i)\mathbf{E}_j + (\mathbf{r}_{ij}\cdot\mathbf{E}_j)\mathbf{E}_i - 5\,\mathbf{r}_{ij}(\mathbf{E}_i\cdot\mathbf{r}_{ij})(\mathbf{E}_j\cdot\mathbf{r}_{ij})\right). \tag{11}$$

Here $\mathbf{r}_{ij}$ is the unit vector from the center of i th particle to j th particle and $\mathbf{r} = |\mathbf{r}_{ij}|$.

We first obtain the electric potential by accounting for the presence of particle:

$$\nabla\cdot(\varepsilon\nabla\varphi) = 0. \tag{12}$$

The boundary conditions on the particle surface are given by

$$\begin{aligned}\varphi_1 &= \varphi_2\\ \varepsilon_c\frac{\partial\varphi_1}{\partial n} &= \varepsilon_p\frac{\partial\varphi_2}{\partial n}\end{aligned} \tag{13}$$

where φ_1 and φ_2 are the electric potential in the liquid and particles, respectively. The electric field is then calculated using the expression

$$\mathbf{E} = -\nabla\phi, \tag{14}$$

Then the Maxwell stress tensor is computed

$$\sigma_M = \varepsilon\mathbf{E}\mathbf{E} - \frac{1}{2}\varepsilon(\mathbf{E}\bullet\mathbf{E})\mathbf{I} \tag{15}$$

The electrostatic force and torque on the i th particle are given by

$$\mathbf{F}_{E,i} = \int_{\Gamma_{P_i}} \sigma_M \times \mathbf{n}_i\, ds \tag{16}$$

$$\mathbf{T}_{E,i} = \int_{\Gamma_{P_i}} (\mathbf{x}-\mathbf{x}_i)\times\sigma_M\cdot\mathbf{n}_i\, ds \tag{17}$$

where $\mathbf{n}_i$ is the unit outer normal on the surface of the i th particles and $\mathbf{x}_i$ is the center of the i th particle.

2.1.2. Dimensionless Equations and Parameters

Equations (1), (2) and (4) are nondimensionalized by assuming that the characteristic length, velocity, time, stress, angular velocity and electric field scales are a, U^*, a/U^*, $\eta U^*/a$, U^*/a and E_0, respectively. The gradient of the electric field is assumed to scale as E_0/L, where L is the distance between the electrodes, which is taken to be the same as the domain width. The nondimensional equations for the liquid, after using the same symbols for the dimensionless variables, are:

$$\mathrm{Re}\left(\frac{\partial\mathbf{u}}{\partial t} + \mathbf{u}\cdot\nabla\mathbf{u}\right) = -\nabla p + \nabla\cdot\sigma \tag{18}$$

$$\nabla \cdot \mathbf{u} = 0 \text{ in } \Omega \backslash \overline{P(t)} \tag{19}$$

where σ is the dimensionless extra stress. If the PD approximation is used for evaluating the electrostatic force, the dimensionless equations for the particles become

$$\frac{d\mathbf{U}}{dt} = \frac{6\pi\eta a^2}{mU^*} \int \left(\frac{-p\mathbf{I} + \sigma}{6\pi} \right) \cdot n\, ds + \frac{4\pi a^4 \varepsilon_0 \varepsilon_c \beta |E_0|^2}{mU^{*2}L} \mathbf{E} \cdot \nabla \mathbf{E} + \frac{3\pi\varepsilon_0\varepsilon_c a^3 \beta^2 |E_0|^2}{4mU^{*2}|\mathbf{r}_{ij}|^4} \mathbf{F}_D \tag{20}$$

$$I_i \frac{d\omega_i}{dt} = \int (\mathbf{x} - \mathbf{x}_i) \times [(-p\mathbf{I} + \sigma) \cdot n]\, dA + \mathbf{T}_{E,i} \tag{21}$$

In terms of the Maxwell stress tensor, they are:

$$\frac{d\mathbf{U}}{dt} = \frac{6\pi\eta a^2}{mU^*} \int (-p\mathbf{I} + \sigma) \cdot \mathbf{n} ds \quad + \frac{\varepsilon E_0^2 a^3}{mU^{*2}} \int \sigma_M \cdot \mathbf{n} ds \tag{22}$$

$$\frac{d\omega}{dt} = \frac{5\eta a^2}{2mU^*} \int (\mathbf{x} - \mathbf{x}_i) \times (-p\mathbf{I} + \sigma) \cdot \mathbf{n} ds \quad + \frac{\varepsilon E_0^2 a^3}{mU^{*2}} \int (\mathbf{x} - \mathbf{x}_i) \times \sigma_M \cdot \mathbf{n} ds. \tag{23}$$

If the PD approximation is used, the above equations contain the following dimensionless parameters:

$$\text{Re} = \frac{\rho_L U^* a}{\eta}, P_1 = \frac{6\pi\eta a^2}{mU^*}, P_2 = \frac{3\pi\varepsilon_0\varepsilon_c \beta^2 a^3 |E_0|^2}{4mU^{*2}}, P_3 = \frac{4\pi\varepsilon_0\varepsilon_c \beta a^4 |E_0|^2}{mU^{*2}L}, h' = \frac{L}{a}. \tag{24}$$

Here Re is the Reynolds number, which determines the relative importance of the fluid inertia and viscous forces, P_1 is the ratio of the viscous and inertia forces, P_2 is the ratio of the electrostatic particle-particle interaction and inertia forces, and P_3 is the ratio of the dielectrophoretic and inertia forces. Another important parameter, which does not appear directly in the above equations, is the solids fraction of the particles; the rheological properties of ER (electrorheological) suspensions depend strongly on the solids fraction.

If the Maxwell stress tensor approach is used, an alternative parameter $P_E = \frac{\varepsilon E_0^2 a^3}{mU^{*2}}$ which is a measure of the electrostatic forces is obtained in place of P_2 and P_3. This parameter is less informative than parameters P_2 and P_3, as it does not separately quantify the particle-particle and dielectrophoretic contributions. In flows for which the applied pressure gradient or the imposed flow velocity is zero, a characteristic velocity $U^* = \frac{2\varepsilon_0\varepsilon_c \beta a^2 |E_0|^2}{3\eta L}$ can be obtained by assuming that the dielectrophoretic force and the viscous drag terms balance each other. For our simulation results, U^* is the maximum particle velocity.

In order to investigate the relative importance of the electrostatic particle-particle and dielectrophoretic forces, another parameter is defined:

$$P_4 = \frac{P_2}{P_3} = \frac{3\beta L}{16a} \tag{25}$$

Equation (25) implies that if $\beta = O(1)$ and $L >> a$, and so $P_4 > 1$, the particle-particle interaction forces will dominate, which is the case in most applications of dielectrophoresis. On the other hand, if $\beta << O(1)$ and $\frac{L}{a} \sim O(1)$, and thus $P_4 < 1$, then the dielectrophoretic force dominates. The latter is the case for the larger sized particles considered in our simulations.

2.2. Finite Element Method

The computational scheme uses a distributed Lagrange multiplier method (DLM) for simulating the motion of rigid particles suspended in a Newtonian fluid [26,27]. In our implementation of the scheme, the fluid-particle system is treated implicitly using a combined weak formulation in which the forces and moments between the particles and fluid cancel. The flow inside the particles is forced to be a rigid body motion by a distribution of Lagrange multipliers. The time integration is performed using the Marchuk-Yanenko operator-splitting technique which is first-order accurate [19,21,26]. The domain is discretized using a tetrahedral mesh where the velocity and pressures fields are discretized using P_2-P_1 interpolations and the electric potential is discretized using P_2 interpolation. The scheme allows us to decouple its four primary difficulties:

1. The incompressibility condition, and the related unknown pressure, This gives rise to a Stokes-like problem for the velocity and pressure distributions, which is solved by using a conjugate gradient method [32,37].
2. The nonlinear advection term. This gives rise to a nonlinear problem for the velocity which is solved by using a least square conjugate gradient algorithm [32,37].
3. The constraint of rigid-body motion in $P_h(t)$, and the related distributed Lagrange multiplier. This step is used to obtain the distributed Lagrange multiplier that enforces rigid body motion inside the particles. This problem is solved by using a conjugate gradient method described in Reference [32,37]. In our implementation of the method we have used an H^1 inner product (see (35)) for obtaining the distributions over the particles, as the discretized velocity is in H^1.
4. The equation for the electric potential. It is solved independently as it is not directly coupled with the momentum and mass conservation equations. Then the electric force is obtained using the Maxwell stress tensor approach [18,20].

2.3. Simulation Domains and Parameters

To test our finite element code, we conducted direct numerical simulations of the motions of two spherical particles in three-dimensional rectangular channels in which the electric field was non-uniform. In one of the domains, two electrodes were placed in one of the sides of the channel and in the second domain the electrodes were mounted on four walls of the channel.

In the first domain, as shown in Figures 2 and 3, two equally-spaced electrodes are embedded in the left wall parallel to the yz-plane. The electrodes are mounted in the middle of the wall such that they are equidistant from the domain centerline. Notice that the electrodes cover only a fraction of the wall area and so the electric field they generate in the domain is non-uniform. We also assume that the electrodes are mounted inside the walls so that they do not affect the fluid boundary conditions on the surface. The width of the electrodes in the y-direction is equal to the domain width which ensures that the electric field does not vary in the y-direction.

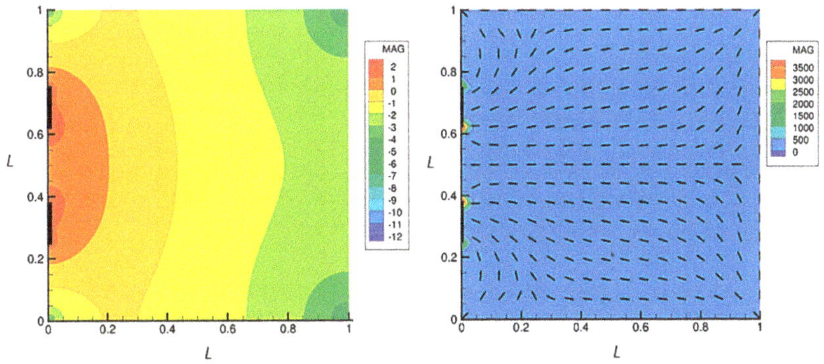

(a) (b)

Figure 2. Electric field on the domain midplane (y = 0.25). (**a**) Isovalues of log $|\mathbf{E}|$. The magnitude of $|\mathbf{E}|$ is maximal near the electrode edges. (**b**) Magnitude and direction of $\nabla \mathbf{E}^2$. The magnitude is shown by isovalues of $|\nabla \mathbf{E}^2|$ and the direction by arrows (the direction of positive DEP force is in this direction and of the negative DEP force in the opposite direction).

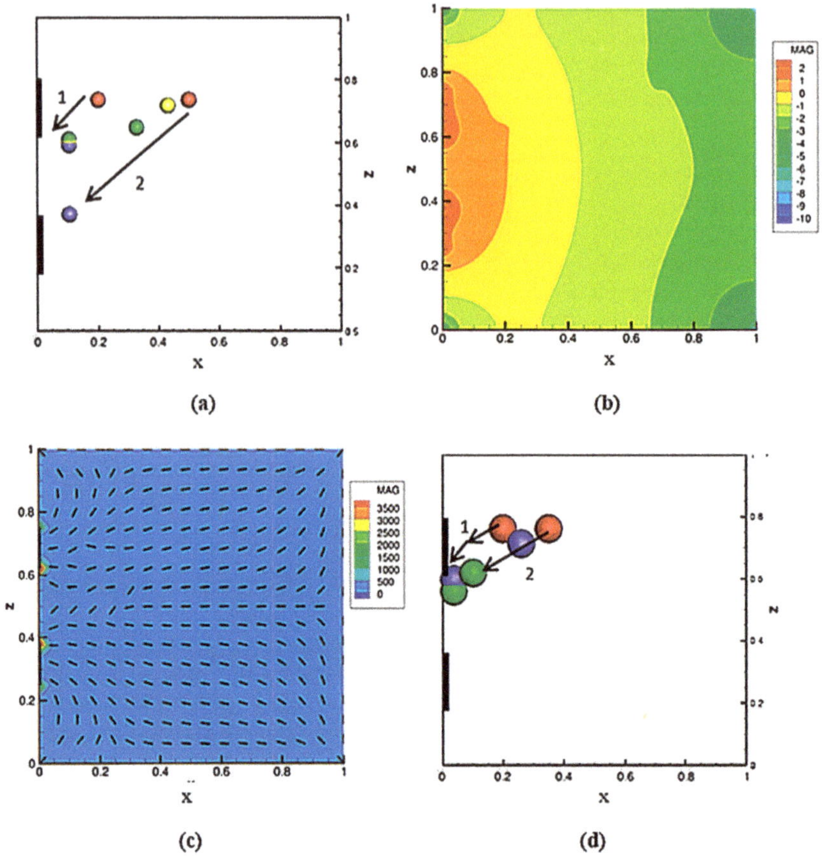

Figure 3. The DEP force induced the motion of two particles in the xz-plane for $\varepsilon_p = 1.2$. (**a**) The diameter of particles is 0.2. The initial positions of the two particles, shown by red circles, are (0.2, 0.25, 0.88) and (0.5, 0.25, 0.88). The final positions are shown using blue circles. Notice that, for clarity, the circles used to show the particles are *smaller* than the actual particle size. (**b**) Isovalues of $\log|\mathbf{E}|$; the particles are at their initial positions. Notice that the electric field is modified from that in Figure 1 because of the presence of particles. (**c**) Magnitude and direction of $\nabla \mathbf{E}^2$; the particles are at (0.11, 0.25, 0.6) and (0.23, 0.25, 0.6). Notice that the direction of the DEP force is no longer symmetric about the domain midplane. (**d**) The diameter of particles is 0.05. The initial positions of the particles, shown by red circles, are (0.2, 0.25, 0.88) and (0.5, 0.25, 0.88). The final positions are shown using blue circles. Notice that in this case both particles collected at the upper electrode.

In the second domain, as shown in Figure 4, there are four electrodes mounted in the four side walls parallel to the yz- and xy-planes. The electrodes are mounted in the middle of the side walls and their depth is equal to the channel depth. The width of the electrodes is one and half the width of the domain sides. This domain can be used to capture a negatively polarized particle at its center and so it is referred to as the dielectrophorectic cage configuration.

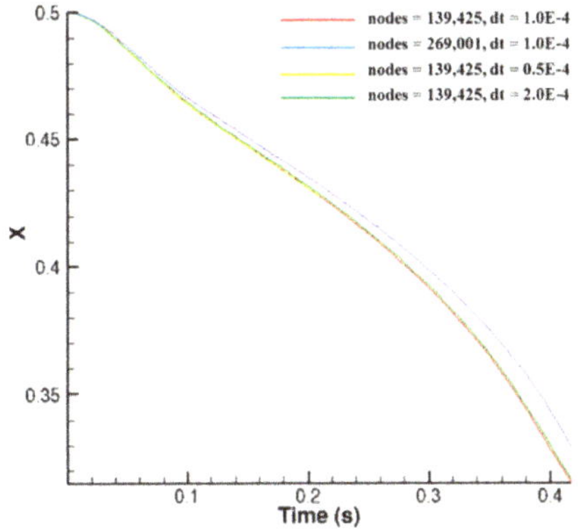

Figure 4. The x-coordinate of the first particle as a function of time for two different mesh resolutions and three different time-step sizes.

In this paper, we will assume that the fluid density ρ_L = 1.0 g/cm^3 and that the particles are neutrally buoyant. The fluid viscosity is η = 0.01 g/(cm·s), and both particles and fluid are non-conducting. The normalized dielectric constant of particles in our simulations is varied and that of the fluid is fixed at one. We will also assume that all of the dimensions reported in this paper are in mm and the time is in seconds. In our simulations, the initial fluid and particle velocities are assumed to be zero. At the domain walls the fluid velocity is assumed to be zero.

As shown in Figure 2, the dimensions of the computational domains in the x-, y- and z-directions are L, $L/2$ and L, respectively. All lengths are nondimensionalized by dividing by L. The particle radius and the distance between the particles are additional characteristic length scales in this problem. Note that the particle dielectric constant has been normalized with respect to the ambient liquid. Thus, a particle with $\varepsilon_p > 1$ is positively polarized and if $\varepsilon_p < 1$ the particle is negatively polarized. The radius of the circles used to represent particles is smaller than the actual particle radius.

3. Results and Discussion

3.1. Domain with the Electrodes Mounted in a Side Wall

The magnitude of the electric field distribution ($|\mathbf{E}|$) on the xz-plane passing through the center of the domain is shown in Figure 2. The field is calculated for the case when the particles are not present and so the dielectric constant does not vary in the domain. Notice that the electric field distribution inside the channel is non-uniform and that the magnitude of $|\mathbf{E}|$ is maximum near the electrode edges and decreases with increasing distance from the electrodes. As discussed earlier in the paper, when particles are placed in a non-uniform electric field they are subjected to both the particle-particle interaction forces and the dielectrophoretic forces. Thus, their transient motions depend on the magnitudes and directions of **E** and $\mathbf{E}\cdot\nabla\mathbf{E}$, and also on the particles positions relative to the electric field direction.

Figure 2b shows that the gradient lines of the electric field magnitude ($\nabla\mathbf{E}^2$) point toward the electrodes, i.e., in the direction normal to the isovalues of $|\mathbf{E}|$. Therefore, the DEP force on a particle with a dielectric constant greater than that of the suspending liquid is directed towards the electrodes; which is in the same direction as the lines of gradient of the electric field magnitude, as indicated by the arrows in Figure 2b. If the dielectric constant of the particle is smaller than that of the liquid, the DEP force is away from the electrodes.

3.2. Motion of Two Particles and Convergence Study

We next describe the transient motion of the two particles in the above non-uniform electric field and analyze the influence of the dipole-dipole and DEP forces. The dimensionless diameter of the particles in Figure 3a is 0.2. The initial positions of the particles are (0.2, 0.25, 0.88) and (0.5, 0.25, 0.88). The dimensionless parameters for the case presented are: Re = 0.652, P_E = 74.7, and P_4 = 0.117. Figure 3 shows the positions of the particles. The positions of the particles are shown at fixed time-intervals and the arrows show the direction of their motion. To track the movement of each particle, we use RYGB color-coding, with red showing the starting position and blue showing the final position. Initially, the particles moved closer because of the particle-particle interaction force, while moving toward the nearest electrode under the action of the dielectrophoretic forces. The particle on the left reached the electrode wall faster, as shown by the yellow circles, as the DEP force acting on it is greater than the particle-particle force because it is close to the upper electrode and the gradient of electric field is larger near the electrodes. The dielectrophoretic force increases as the distance between the particles and the electrode decreases, and the particle-particle force becomes relatively less in magnitude. The left particle is thus collected at the electrode much more quickly. The second set of particles initially moved slowly as the dielectrophoretic force acting on it was smaller, but its speed increased as it moved closer to the electrode. Notice that the electric field intensity and the magnitude and direction of $\nabla \mathbf{E}^2$ are modified by the particles (Figure 3c,d). The DEP force lines are no longer symmetric about the domain midplane, and the DEP force on the right particle is towards the lower electrode. The right particle, therefore, collected at the lower electrode. The particle-particle interaction force in this case is weaker than the DEP force which is expected since P_4 = 0.117. In Figure 3d, we present the results for the case where the particle diameter is 0.05, and P_4 = 0.468. Notice that in this case both particles collected at the upper electrode. This is due to the fact that the DEP force in this case is not stronger than the particle-particle force since the particle size is smaller.

Convergence Study

A convergence study is conducted to show that the numerical results obtained using the MST method converge when the time-step size is reduced and when the mesh is refined. To show convergence with the mesh refinement, we performed simulation on two mesh sizes. The first contained 139,425 nodes and the second 269,001 nodes. The times-step size used for these simulations is 1.0×10^{-4} s. To show convergence with the time-step size, simulations were performed for the time-step sizes of 2×10^{-4} s, and 1×10^{-4} s and 0.5×10^{-4} s. The number of nodes in these simulations is 139,425.

Figure 4 shows the x-position of the first particle as a function of time for two different mesh resolutions and three different time-step sizes. The figure shows that the trajectory of the particle remains virtually unchanged when the time step is reduced and also the mesh is refined. We may, therefore, conclude that the numerical results are converged and so are independent of the mesh refinement and the time step used.

3.3. Motion in a Dielectrophoretic Cage

As noted earlier, a dielectrophoretic cage is formed by placing electrodes in the four sides of a square-shaped domain as shown in Figure 5. The same voltage is applied to the opposite sides and the voltage applied to the adjacent sides is of the opposite sign. This device is of practical interest because it provides a way to trap one or more particles at the center of the device in a contactless fashion. The magnitude of the electric field ($|\mathbf{E}|$) on the *xz*-plane passing through the middle of the domain is shown in Figure 5. Notice that there is a local minimum of $|\mathbf{E}|$ at the center of the domain, and that the field magnitude near the center increases with increasing distance from the center. The figure also shows that the electric field inside the cage is non-uniform, and that its gradient near the domain center is non-zero, except at the center itself where it is zero. Thus, the DEP force acting on a particle at the center is zero. The gradient of electric field magnitude points radially outward from the center and towards the electrodes (see Figure 5b). Therefore, if the dielectric constant of a particle is smaller than that of the liquid, the DEP force on the particle near the center will be towards the center, i.e., in the direction opposite to the gradient lines of the electric field magnitude. This implies that if the particle drifts away from the center, a restoring force acts on it which brings it back to the center. On the other hand, if the dielectric constant of the particle is greater than that of the liquid, the direction of the electric force is away from the center and so the particle moves away from the center and is carried to an electrode edge.

We next consider the motion of two particles in a dielectrophoretic cage, under the influence of positive and negative dielectrophoresis. For positive dielectrophoresis, $\varepsilon_p > \varepsilon_c = 1$, and so the DEP force is in the direction of the force lines shown in Figure 5b, and for negative dielectrophoresis, $\varepsilon_p < \varepsilon_c = 1$, the DEP force is in the direction of the force lines shown in Figure 4b.

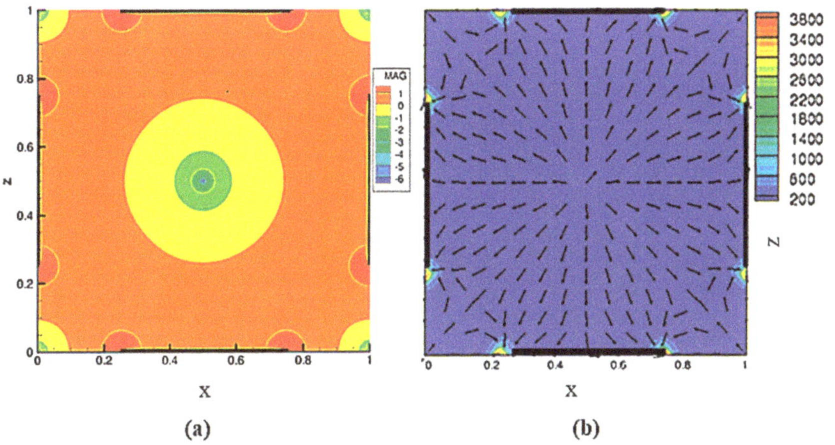

Figure 5. Electric field (log | **E** |) and $\nabla \mathbf{E}^2$ distributions in a DEP cage. The dimensionless parameters are: Re = 0.113, P_E = 3080, and P_4 = 0.668. (**a**) Isovalues of log | **E** | at y = 0.25, i.e., the domain mid-plane. Electric field does not vary in the y-direction. The magnitude of | **E** | is maximal near the electrode edges and decreases with increasing distance from the electrodes. The minimum of | **E** | is at the center. (**b**) The magnitude and direction of $\nabla \mathbf{E}^2$ to which the DEP force is proportional. Arrows indicate the direction of positive DEP force (for negative DEP, the direction is the opposite) and the isovalues levels give the magnitude.

3.3.1. Positive Dielectrophoresis $(\varepsilon_p = 1.2)$

We first consider the case of positive dielectrophoresis, and so the dielectric constant of the particles ($\varepsilon_p = 1.2$) is assumed to be greater than that of the suspending fluid $\varepsilon_c = 1.0$. Figure 6 shows the positions of two particles for four different initial positions. The particles position at selected times are shown using a color-coded RYGB scheme, with red showing the starting position and blue showing the final position. The figure shows that the two particles are collected at a nearby electrode in all four cases.

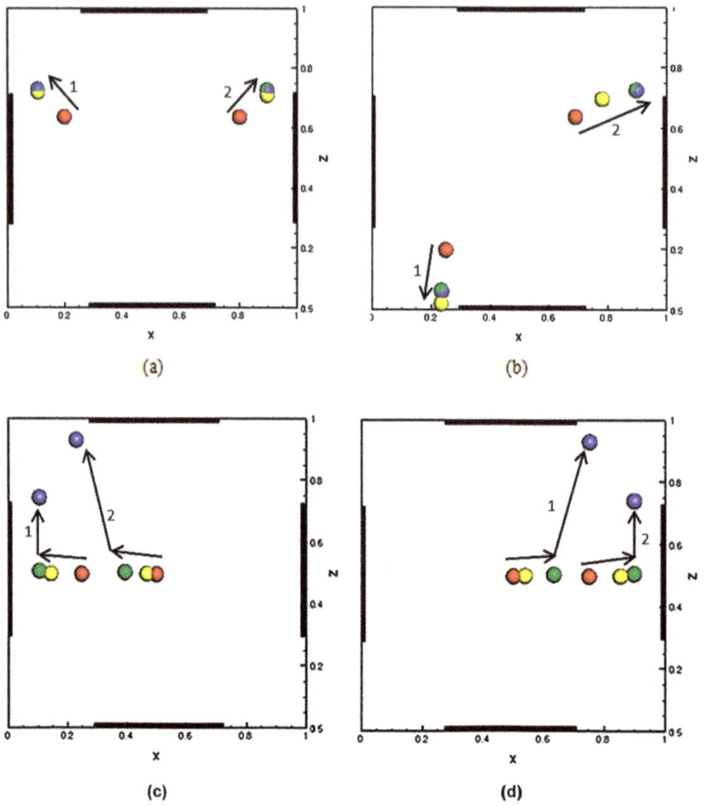

Figure 6. The DEP force-induced motion of two particles in the xz-plane under positive dielectrophoresis. The diameter of particles is 0.2. The initial positions of the two particles, shown by red circles, are: (**a**) (0.2, 0.25, 0.65) and (0.8, 0.25, 0.65); (**b**) (0.25, 0.25, 0.2) and (0.5, 0.25, 0.65); (**c**) (0.25, 0.25, 0.50) and (0.5, 0.25, 0.50); (**d**) (0.75, 0.25, 0.50) and (0.50, 0.25, 0.50). The dimensionless parameters are: Re = 0.652, P_E = 74.7, and P_4 = 0.117.

Notice that in most cases particles are collected at the electrode edge that is nearest to them. However, a particle near the domain center can be pulled away from the nearest electrode by the particle-particle interaction force so that it is collected at an electrode edge that is farther away. For instance, in Figure 6c,d the final position of the particle at the center of the domain is determined by the initial position of the second particle which is near the center. The particle at the center is not subjected to a DEP force because the gradient of the electric field at the center is zero, and so if the second particle was not present it would remain at the center. In Figure 6c,d, the second particle pulls the particle at the center in its direction, and after it moves away from the center the DEP force also starts to act, and the DEP and particle-particles forces collectively determine its subsequent trajectory. The particle which was initially at the center is collected at the upper electrode because the particle-particle force in this case is smaller than the DEP force; P_4 = 0.117. The presence of particles modifies the DEP force lines, as was the case in Figure 3, and the DEP force drives the particle to the upper electrode. As discussed in Section 3.2, when the distance between the electrodes is comparable to the particle size, the tendency to form chains is weaker, especially near the electrodes, and so particles move individually.

These results are similar to the results in Reference [35] where the parameter P_4 was reduced by reducing β (which was varied independently by changing the frequency). Specifically, it was shown that since the particle-particle interaction force depends on the second power of β, and therefore becomes negligible compared to the DEP force, which itself depends linearly on β when β is small. This diminished the role of particle-particle interaction forces and, as a result, yeast particles moved individually, without forming chains, because of the DEP force. Here, on the other hand, the parameter P_4 has been reduced by increasing the particle size.

3.3.2. Negative Dielectrophoresis $(\varepsilon_p = 0.2)$

In this subsection, we consider the case where the dielectric constant of the particles $(\varepsilon_p = 0.2)$ is smaller than that of the suspending fluid $\varepsilon_c = 1.0$ and so particles undergo negative dielectrophoresis. For the four test cases shown in Figure 7, all other parameters including the domain size and the initial positions of the particles, are the same as for the case of the positive dielectrophoresis described in the previous subsection.

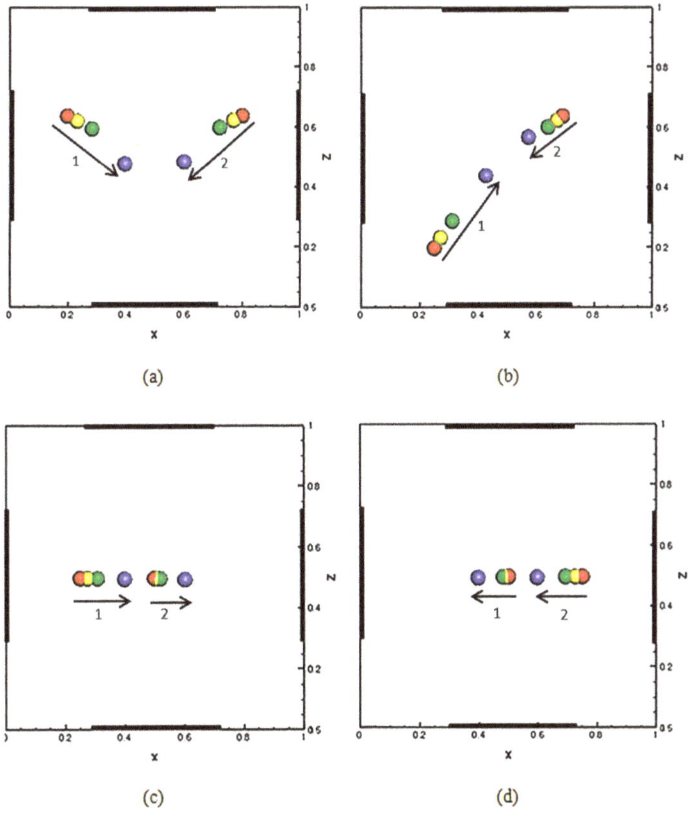

Figure 7. The DEP force-induced motion of two particles in the *xz*-plane under negative dielectrophoresis. The diameter of particles is 0.1. The initial positions of the particles, shown by red circles, are: (**a**) (0.2, 0.25, 0.65) and (0.8, 0.25, 0.65); (**b**) (0.25, 0.25, 0.2) and (0.5, 0.25, 0.65); (**c**) (0.25, 0.25, 0.50) and (0.5, 0.25, 0.50; (**d**) (0.75, 0.25, 0.50) and (0.50, 0.25, 0.50). The dimensionless parameters are: Re = 0.0562, P_E = 1670, and P_4 = 0.682.

From Figure 7, we note that in all four cases the particles moved towards the domain center where the electric field magnitude is locally minimal (see Figure 5a). This also shows that the particle pair is stable at the domain center and that if the position is perturbed away from the center a restoring force acts which brings it back to the center. Thus, a dielectrophoretic cage can be used to position and hold negatively polarized particles at its center. Also, notice that the particles' motion was in the opposite direction to the gradient line of the electric field magnitude ($\nabla \mathbf{E}^2$) as shown in Figure 4b. In fact, even after they are collected at the domain center, the particles maintained their orientations parallel to the gradient line along which they traveled. At the center, the gradient lines of ($\nabla \mathbf{E}^2$) emanate approximately radially outward, and so the particle pair is not subjected to a torque and its final orientation depends only on the initial orientation. Also, note that if a particle at the center is rotated, it is not subjected to a restoring torque and so it does not come back to its original orientation. The DEP force can only bring and hold circular particles at its center but cannot keep their orientations fixed.

3.4. Comparison with the Point-Dipole Approach

Finally, we compare the trajectory of a single particle in a DEP cage obtained using the MST method with that obtained using the point dipole (PD) method. We remind the reader that the PD method assumes that the presence of a particle does not alter the electric field and that the electric force on the particle can be obtained by using the value of $\nabla \mathbf{E}^2$ at its center. However, this is not the case for the results presented in Figures 5a and 8 which show that the isovalues of log $|\mathbf{E}|$ are altered by the presence of a particle, and so the electric force given by the two methods is different. The difference between the forces given by the two methods is greater when the particle size is comparable to the length scale over which the electric field varies and the dielectric mismatch is larger.

In Figure 9, the particle trajectory and velocity computed using the two methods are shown as a function of time. The particle moves from its initial position (0.35, 0.25, 0.40) near the diagonal of the DEP cage to the bottom edge of the left electrode. Notice that although the final position given by the two force methods are approximately the same, the computed trajectories are different. For the point dipole method, the time taken to reach the equilibrium position was 0.15 s and for the MST method it was 0.13 s, implying that the point dipole method underestimates the electric force. This is consistent with the results reported in Reference [18], where the force acting on a fixed particle was computed using the MST and point dipole methods, i.e., the fluid and particle were not moving. It was noted that along the cage diagonal the PD method underestimates the force, and that the difference is larger when the particle size is larger. However, the PD method overestimates the force if the particle is displaced along the x- or y-direction from the center (also see [22]).

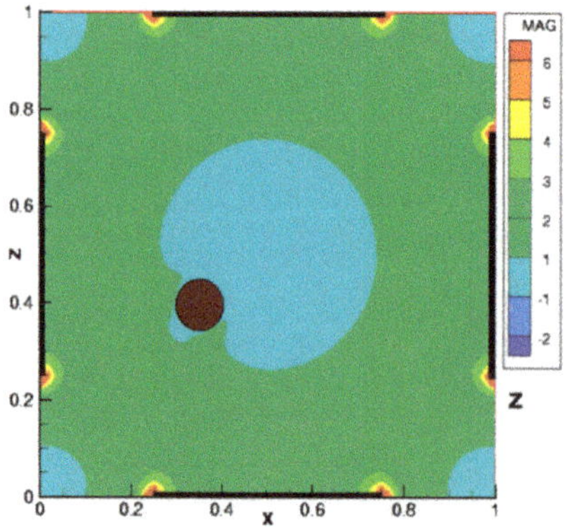

Figure 8. Isovalues of log $|\mathbf{E}|$ at the domain mid-plane, $y = 0.25$, with a particle present (a = 0.05 and $\varepsilon_p = 1.5$). The dimensionless parameters are: Re = 0.113, $P_E = 3080$, and $P_4 = 0.668$.

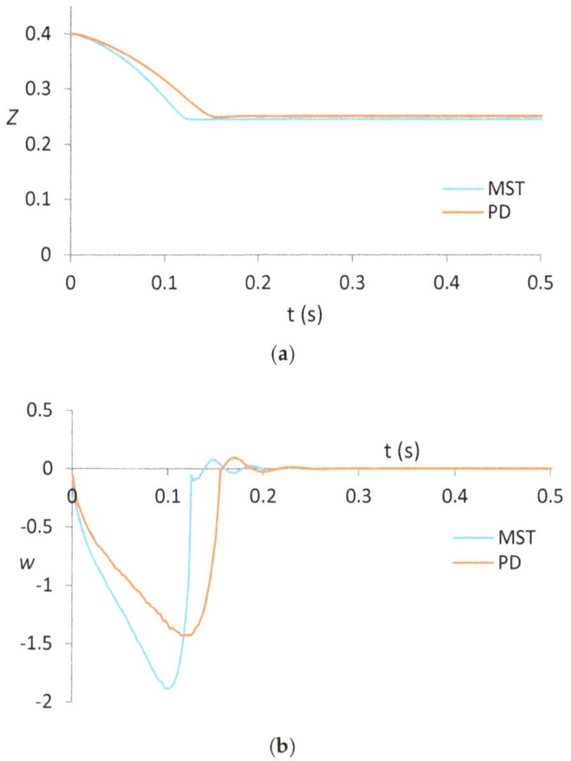

Figure 9. The z-components of particle position (z) and velocity (w) for the PD and MST methods are shown as functions of time. a = 0.05 and $\varepsilon_p = 1.5$ and the initial particle position is (0.35, 0.25, 0.40). (**a**) z and (**b**) w. The dimensionless parameters are: Re = 0.113, P_E = 3080, and P_4 = 0.668.

In Figures 10 and 11, the particle positions and velocities obtained using the PD and MST methods are compared for the cases where the two particles are released away from their equilibrium positions in the cage. In Figure 10, the particles collected at an electrode edge where the magnitude of the electric field gradient is maximal. Notice that the final particle positions are slightly different for the two methods. This is due to the fact that the PD method does not account for the electric field modification caused by the presence of particles. Also, since the electric field gradient near the electrodes is larger, the differences in the trajectories for the two methods become larger as the particles move closer to the electrodes.

In Figure 11, the differences in the trajectories obtained using the PD and MST methods are relatively smaller. In this case, the particles undergo negative dielectrophoresis and move closer to the cage center where the magnitude of the electric field gradient is smaller (see Figure 8). These results show that in the regions where the gradient of the electric field intensity is larger or when the particle radius is comparable to the length scale over which the electric field varies, the difference between the trajectories obtained for the MST and the point-dipole methods is larger.

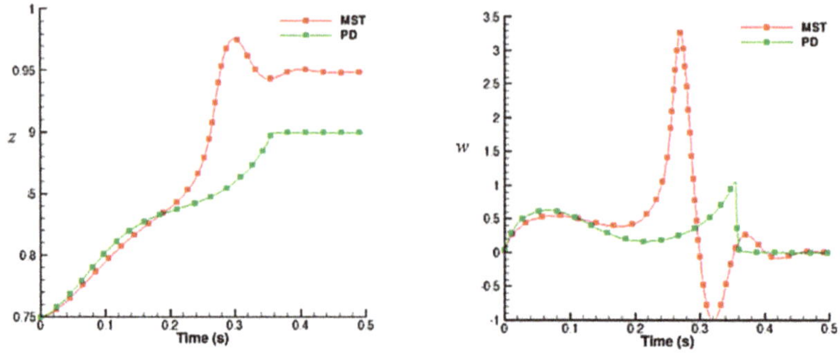

(a) (b)

Figure 10. The *z*-components of the second particle position (*z*) and velocity (*w*) for the PD and MST methods are shown as functions of time. $a = 0.1$ and $\varepsilon_p = 1.5$. The initial positions of the particles are (0.25, 0.25, 0.75) and (0.75, 0.25, 0.75). (**a**) z, (**b**) *w*. The dimensionless parameters are the same as in Figure 5.

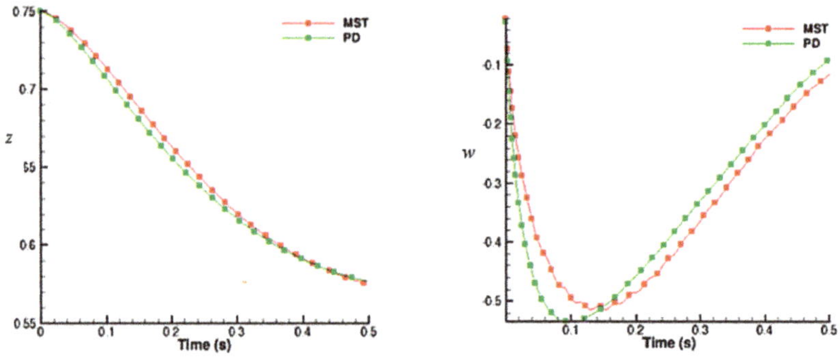

(a) (b)

Figure 11. The *z*-components of the second particle position (*z*) and velocity (*w*) for the PD and MST methods are shown as functions of time. $a = 0.1$ and $\varepsilon_p = 0.7$. The initial positions for the particles are (0.25, 0.25, 0.75) and (0.75, 0.25, 0.75). (**a**) z, (**b**) *w*. The dimensionless parameters are the same as in Figure 6.

4. Conclusions

A finite element scheme based on the distributed Lagrange multiplier method is used to study the dynamical behavior of particles in a dielectrophoretic cage. The Maxwell stress tensor method is used for computing the electric forces acting on the particles, and the Marchuk-Yanenko operator-splitting technique is used to discretize in time. It is shown that a dielectrophoretic cage can be used to trap and hold negatively polarized particles at its center. If the particle moves away from the center of the cage, a resorting force acts on the particle towards the center. The cage also allows two particles to be trapped simultaneously at the center. The orientation of a trapped particle pair at the center depends on the initial positions of the particles, and in this sense, the pair orientation at the center is not fixed. Positively polarized particles, on the other hand, are trapped at the electrode edges depending on their initial locations. The particle trajectories obtained using the MST and point-dipole methods differ, but the final steady positions of the particles are approximately the same. The ratio of the particle-particle interaction and dielectrophoretic forces, P_4, decreases with increasing particle size which diminishes the tendency of particles to form chains, especially when they are close to the electrodes. Also, when the spacing between the electrodes is comparable to the particle size, instead of forming chains on the same electrode, particles collect at different electrodes. This is a consequence of the modification of the electric field due to the presence of particles, which is greater for larger particles.

Author Contributions: Conceptualization, E.A., M.J., P.S.; Methodology, E.A., M.J., P.S.; Software, P.S.; Validation, E.A., M.J., P.S.; Formal Analysis, E.A., M.J., P.S.; Investigation, E.A., M.J., P.S.; Resources, P.S.; Data Curation, E.A., M.J., P.S.; Writing-Original Draft Preparation, E.A., M.J., P.S.; Writing-Review & Editing, E.A., M.J., P.S.; Visualization, E.A., M.J., P.S.; Supervision, P.S.; Project Administration, P.S.; Funding Acquisition, P.S.

Funding: The authors thank the financial support from the National Science Foundation (CBET-1067004).

Conflicts of Interest: The authors declare no conflict of interest. The founding sponsors had no role in the design of the study; in the collection, analyses, or interpretation of data; in the writing of the manuscript, and in the decision to publish the results.

References

1. Washizu, M.; Kurosawa, O.; Arai, I.; Suzuki, S.; Shimamoto, N. Applications of electrostatic stretch-and-positioning of DNA. *IEEE Trans. Ind. Appl.* **1995**, *30*, 835. [CrossRef]
2. Pohl, H.A. *Dielectrophoresis: The Behavior of Neutral Matter in Non-Uniform Electric Fields*; Cambridge University Press: Cambridge, NY, USA, 1978.
3. Hughes, M.P.; Morgan, H.; Rixon, J.F. Measuring the dielectric properties of herpes simplex virus type 1 virions with dielectrophoresis. *Biochim. Biophys. Acta* **2002**, *1571*, 1–8. [CrossRef]
4. Becker, F.F.; Wang, X.; Huang, Y.; Pethig, R.; Vykoukal, J.; Gascoyne, P.R.C. The removal of human leukemia cells from blood using interdigitated microelectrodes. *J. Phys. D Appl. Phys.* **2002**, *27*, 2659. [CrossRef]
5. Becker, F.F.; Wang, X.B.; Huang, Y.; Pethig, R.; Vykoukal, J.; Gascoyne, P.R.C. Separation of Human Breast Cancer Cells From Blood by Differential Dielectric Affinity. *Proc. Natl. Acad. Sci. USA* **1995**, *92*, 860. [CrossRef] [PubMed]
6. Markx, G.H.; Dyda, P.A.; Pethig, R. Dielectrophoretic separation of bacteria using a conductivity gradient. *J. Biotechnol.* **1996**, *51*, 175. [CrossRef]
7. Hughes, M.P.; Morgan, H. Dielectrophoretic trapping of single sub-micrometre scale bioparticles. *J. Phys. D Appl. Phys.* **1998**, *31*, 2205. [CrossRef]
8. Voldman, J.; Braff, R.A.; Toner, M.; Gray, M.L.; Schmidt, M.A. Holding Forces of Single-Particle Dielectrophoretic Traps. *Biophys. J.* **2002**, *80*, 531. [CrossRef]
9. Green, N.G.; Morgan, H.; Milner, J.J. Manipulation and trapping of sub-micron bioparticles using dielectrophoresis. *J. Biochem. Biophys. Methods* **1997**, *35*, 89–102. [CrossRef]
10. Voldman, J.; Toner, M.; Gray, M.L.; Schmidt, M.A. Design and analysis of extruded quadrupolar dielectrophoretic traps. *J. Electrostat.* **2003**, *57*, 69–90. [CrossRef]
11. Pethig, R.; Huang, Y.; Wang, X.B.; Burt, J.B.H. Positive and negative dielectrophoretic collection of colloidal particles using interdigitated castellated microelectrodes. *J. Phys. D Appl. Phys.* **1992**, *25*, 881–888. [CrossRef]
12. Green, N.G.; Morgan, H. Dielectrophoresis of submicrometer latex spheres. 1. Experimental Results. *J. Phys. Chem. B* **1999**, *103*, 41–50. [CrossRef]
13. Hughes, M.P.; Morgan, H. Measurement of bacterial flagellar thrust by negative dielectrophoresis. *Biotechnol. Prog.* **1999**, *15*, 245–249. [CrossRef] [PubMed]
14. Jones, T.B. *Electromechanics of Particles*; Cambridge University Press: Cambridge, NY, USA, 1995.
15. Nedelcu, S.; Watson, J.H.P. Size separation of DNA molecules by pulsed electric field dielectrophoresis. *J. Phys. D Appl. Phys.* **2004**, *37*, 2197. [CrossRef]
16. Washizu, M.; Jones, T.B. Generalized multipolar dielectrophoretic force and electrorotational torque calculation. *J. Electrostat.* **1996**, *38*, 199. [CrossRef]
17. Wang, X.; Wang, X.B.; Gascoyne, P.R. General expressions for dielectrophoretic force and electrorotational torque derived using the Maxwell stress tensor method. *J. Electrostat.* **1997**, *39*, 277. [CrossRef]
18. Singh, P.; Aubry, N. Trapping force on a finite-sized particle in a dielectrophoretic cage. *Phys. Rev. E* **2005**, *72*, 016602. [CrossRef] [PubMed]

19. Singh, P.; Aubry, N. Particle Separation Using Dielectrophoresis. In Proceedings of the ASME Annual Meeting, Anaheim, CA, USA, 13–19 November 2004.
20. Aubry, N.; Singh, P. Control of electrostatic particle-particle interactions in dielectrophoresis. *Europhys. Lett.* **2006**, *74*, 623–629. [CrossRef]
21. Bonnecaze, R.T.; Brady, J.F. Dynamic simulation of an electrorheological fluid. *J. Chem. Phys.* **1992**, *96*, 2183. [CrossRef]
22. Hossan, M.R.; Dillon, R.; Dutta, P. Hybrid immersed interface-immersed boundary methods for AC dielectrophoresis. *J. Comput. Phys.* **2014**, *270*, 640–659. [CrossRef]
23. Hossan, M.R.; Dillon, R.; Roy, A.K.; Dutta, P. Modeling and simulation of dielectrophoretic particle–particle interactions and assembly. *J. Colloid Interface Sci.* **2013**, *394*, 619–629. [CrossRef] [PubMed]
24. Hossan, M.R.; Gopmandal, P.P.; Dillon, R.; Dutta, P. A comprehensive numerical investigation of DC dielectrophoretic particle-particle interactions and assembly. *Colloid Surf. A* **2016**, *506*, 127–137. [CrossRef]
25. Ai, Y.; Qian, S. DC dielectrophoretic particle–particle interactions and their relative motions. *J. Colloid Interface Sci.* **2010**, *346*, 448–454. [CrossRef] [PubMed]
26. Ai, Y.; Zeng, Z.; Qian, S. Direct numerical simulation of AC dielectrophoretic particle-particle interactive motions. *J. Colloid Interface Sci.* **2014**, *417*, 72–79. [CrossRef] [PubMed]
27. House, D.L.; Luo, H.; Chang, S. Numerical study on dielectrophoretic chaining of two ellipsoidal particles. *J. Colloid Interface Sci.* **2012**, *374*, 141–149. [CrossRef] [PubMed]
28. Moncada-Hernandez, H.; Nagler, E.; Minerick, A.R. Theoretical and experimental examination of particle-particle interaction effects on induced dipole moments and dielectrophoretic responses of multiple particle chains. *Electrophoresis* **2014**, *35*, 1803–1813. [CrossRef] [PubMed]
29. Knoerzer, M.; Szydzik, C.; Tovar-Lopez, F.J.; Tang, X.; Mitchell, A.; Khoshmanesh, K. Dynamic drag force based on iterative density mapping: A new numerical tool for three-dimensional analysis of particle trajectories in a dielectrophoretic system. *Electrophoresis* **2016**, *37*, 645–657. [CrossRef] [PubMed]
30. Green, N.G.; Ramos, A.; Morgan, H. Numerical solution of the dielectrophoretic and traveling wave forces for interdigitated electrode arrays using the finite element method. *J. Electrostat.* **2002**, *56*, 235–254. [CrossRef]
31. Li, H.; Bashir, R. On the design and optimization of microfluidic dielectrophoretic devices: A dynamic simulation study. *Biomed. Microdevices* **2002**, *6*, 289–295. [CrossRef] [PubMed]
32. Glowinski, R.; Pan, T.W.; Hesla, T.I.; Joseph, D.D. A distributed Lagrange multiplier/fictitious domain method for particulate flows. *Int. J. Multiph. Flow* **1999**, *25*, 755. [CrossRef]
33. Kadaksham, J.; Singh, P.; Aubry, N. Dynamics of electrorheological suspensions subjected to spatially non-uniform electric fields. *J. Fluids Eng.* **2004**, *120*, 170. [CrossRef]
34. Kadaksham, J.; Singh, P.; Aubry, N. Dielectrophoresis of nanoparticles. *Electrophoresis* **2004**, *25*, 3625. [CrossRef] [PubMed]
35. Kadaksham, J.; Singh, P.; Aubry, N. Dielectrophoresis induced clustering regimes of viable yeast cells. *Electrophoresis* **2005**, *26*, 3738–3744. [CrossRef] [PubMed]
36. Kadaksham, J.; Singh, P.; Aubry, N. Manipulation of particles using dielectrophoresis. *Mech. Res. Commun.* **2006**, *33*, 108–122. [CrossRef]
37. Singh, P.; Glowinski, R.; Pan, T.W.; Hesla, T.I.; Joseph, D.D. A distributed Lagrange multiplier/fictitious domain method for viscoelastic particulate flows. *J. Non-Newton. Fluid Mech.* **2000**, *91*, 165–188. [CrossRef]

fluids

MDPI

Article

Encapsulation of Droplets Using Cusp Formation behind a Drop Rising in a Non-Newtonian Fluid

Raphaël Poryles * and Roberto Zenit *

Instituto de Investigaciones en Materiales, Universidad Nacional Autónoma de México, Mexico DF 04510, Mexico

* Correspondence: raphael.poryles@yahoo.fr (R.P.); zenit@unam.mx (R.Z.)

Received: 4 June 2018; Accepted: 27 July 2018; Published: 1 August 2018

Abstract: The rising of a Newtonian oil drop in a non-Newtonian viscous solution is studied experimentally. In this case, the shape of the ascending drop is strongly affected by the viscoelastic and shear-thinning properties of the surrounding liquid. We found that the so-called velocity discontinuity phenomena is observed for drops larger than a certain critical size. Beyond the critical velocity, the formation of a long tail is observed, from which small droplets are continuously emitted. We determined that the fragmentation of the tail results mainly from the effect of capillary effects. We explore the idea of using this configuration as a new encapsulation technique, where the size and frequency of droplets are directly related to the volume of the main rising drop, for the particular pair of fluids used. These experimental results could lead to other investigations, which could help to predict the droplet formation process by tuning the two fluids' properties, and adjusting only the volume of the main drop.

Keywords: drop; cusp instability; encapsulation

1. Introduction

The problem of encapsulating droplets of fluid has important implications in the fields of bioengineering and medical research, for instance to encapsulate cells [1]. With the development of microfluidics and lab-on-chip technology to perform analysis on different fluids, the dynamics and size of such droplets have to be well controlled [2–4]. Several techniques have been used to perform such encapsulation, for instance using a T-junctions device [5,6]. To be able to perform such encapsulation at a larger scale in a controlled matter still remains to be achieved.

Here, we study a new alternative technique to encapsulate oil drops by using the non-Newtonian properties of the surrounding liquid. In the case of an object rising or falling in a non-Newtonian fluid, new and unexpected phenomena appear in comparison with the Newtonian case. The flow surrounding the object can be highly modified, due to the viscoelastic properties of the fluids [7–13].

It has been observed that when a bubble or a drop moves, as a result of gravity, in a viscoelastic shear-thinning fluid, a velocity discontinuity phenomenon appears [14–19]. When the bubble or drop reaches a certain critical size, its terminal speed increases sharply. The sudden increase has not been understood fully until recently [17]: a combination of effects has to occur simultaneously. First the viscoelastic nature of the outside liquid induces a change in shape of the drop or bubble, forming a characteristic cusped shape [14,15,19–21]; resulting from this change of shape, the drag coefficient of the object is reduced. Consequently, the speed increases, which, given the shear-thinning properties, will cause the viscosity around the object to decrease, causing an even further rising velocity increase. This set of features was recently discussed by [17] in detail. Recent numerical simulations have helped to clarify the connection among these behaviours [22].

Furthermore, when the velocity discontinuity appears, a negative wake behind the object is also detected. The flow field behind the bubble reverses direction. This phenomenon was first observed by [23] and was directly related to the appearance of the velocity discontinuity by [14,19,22].

More interestingly, for the case of a drop, a recent study by [20] found that a long thin tail formed from the cusp at the rear edge of the drop. This long tail became unstable, fragmenting into small droplets that were then left behind the main drop. They observed the tail formation and fragmentation when the external fluid had non-Newtonian properties. Since they conducted experiments with both Newtonian (as in the present case) and non-Newtonian drops, it was possible to determine that viscoelastic drops formed much longer and thicker tails than the Newtonian case. It is important to note that the fragmentation of the tail was not analyzed in detail. This is, in fact, one of the objectives of the present study. The fragmentation of fluid filaments has been extensively studied in the literature under different flow conditions [24,25]. For the case of viscoelastic filament, the breakup has been shown to lead to the formation of bead-on-a-string [26,27]. In most cases, the fragmentation results from capillary-driven instabilities similar to the Rayleigh–Plateau instability [24,25]. Some studies have addressed the case of a Newtonian jet flowing from a nozzle in a viscoelastic fluid [28–30]. In those experiments, the jet fragmenting is controlled by the injection flow, while, in the present case, the filament is created directly by the rising drop.

In this article, we present results on the formation of droplets behind an oil drop (Newtonian) rising in a water/glycerol-polyacrylamide solution (non-Newtonian). First, we present the experimental set-up and a characterization of the fluids we used. Then, we present the experimental observations, and the different regimes of breakup that were observed. Finally, we discuss different aspects of the droplets formation, by relating the velocity and volume of the main drop, the size of the tail appearing behind the drop and the size and frequency of formation of the droplets. These results, obtained only for a single pair of Newtonian/non-Newtonian fluids, could lead to a more extensive investigation with which this phenomenon could be fully understood.

2. Experimental Set-Up and Test Fluids

The experimental setup consists of a vertical glass column of a height 60 cm (Figure 1). This column is square based with a side width of 6 cm. This column is wide enough so that wall effects are negligible on the drop dynamics, as the velocity field decreases rapidly with the distance to the drop. The setup is filled with a non-Newtonian water-polyacrylamide solution. Alimentary corn oil is injected at the bottom of the column using a plastic syringe with a volume capacity of 5 mL. The set-up is backlit with a LED panel. A fast speed camera (SpeedSence, Phantom) films the rising at a frequency of 200 frames per seconds with a resolution 1632×1200 during 10 s. The camera is placed at mid height of the column and films a zone of about 12 cm high. At this point, the drop moves at its terminal speed. The scale ratio of the images is of 110 pixels per centimeter, and we measure the diameter, height and position of the drop with a precision of about two pixels, so the error is estimated to be smaller than 5 percent for the drop velocity and volume (see Section 3). The error bars have been reported on the different figures.

The properties of the two fluids used are presented in Figure 2a. The drop consists in an alimentary corn oil, characterized as Newtonian. For the surrounding fluid, we used a solution composed of a 49.75% weight solution of water and glycerol each and 0.5% of industrial polyacrylamide (PAAM, Separan) which are long chains of polymer. This high concentration ensures that the non-Newtonian behaviour of the fluid will be important (shear-thinning and viscoelasticity). The density was measured using a flask with a precise volume of 25 mL, which was filled with the different fluids and weighed.

Both fluids were characterized using a rheometer (HR, TA Instruments), and we performed two types of tests. We used a plane–plane geometry with a gap of 1 mm, at a fixed temperature of 25 °C and we measured the viscosity, varying the shear-rate. The shear rate $\dot{\gamma}$ is varied from 0.01 to 100 s^{-1} for the polymer solution and from 0.1 to 100 s^{-1} for the oil with five points per decade (for the oil, the rheometer does not have sufficient accuracy for $\dot{\gamma} < 0.01\ s^{-1}$ because of the low viscosity), with an averaging time of 30 s for each point, and with back and forth variation for reproducibility (going from low shear rate to high, and then reverse). The results are presented in Figure 2b. We observe that the corn oil is Newtonian, with a viscosity $\eta_{oil} \approx 0.06$ Pa.s. The non-Newtonian fluid, shows a shear-thinning behaviour, as the viscosity η_{PAAM} (red squares) decreases with the shear rate $\dot{\gamma}$. In our case, the shear-rates when the drop ascends in a range between $\dot{\gamma} = 0.22\ s^{-1}$ and $\dot{\gamma} = 0.42\ s^{-1}$. In this zone, the viscosity follows a power law: $\eta = K\dot{\gamma}^{n-1}$, where $n = 0.87$ and $K = 1.10$ as represented in Figure 2b. This type of decrease is typical for polyacryalimde solutions [31,32]. This exponent n being close to 1 (Newtonian behaviour for $n = 1$) indicates that the shear-thinning is insignificant in our experiment. The important decrease in viscosity η will appear for shear-rates $\dot{\gamma}$ higher than those relevant here.

For the non-Newtonian solution, an oscillatory test was also performed. A deformation of 3% was imposed, and the frequency of oscillation ω was varied from 0.06 to 100 rad·s^{-1} during two periods for each point. With these measurements, the elastic G' (empty symbols) and viscous modulus G'' (filled symbols) are obtained. We observe a viscoelastic behaviour, where the elastic property is dominant at a low shear rate. This is in agreement with what has already been observed for such polymer solutions [32]. The relaxation time can be approximated as the time where the two moduli are equal and represents the typical time where the non-Newtonian fluid goes from an elastic solid behaviour to a viscous fluid. This is represented in Figure 2c as $\omega_r = 25$ rad·s^{-1}, and we estimate the relaxation time $\tau_r = 2\pi/\omega_r = 0.25$ s.

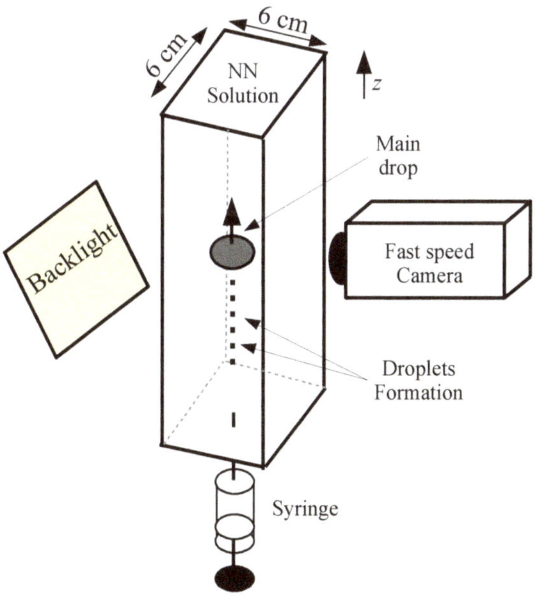

Figure 1. Scheme of the experimental set-up. In a vertical glass column with a square base of 6 cm side width and a height of 60 cm, we place a non-Newtonian fluid. An oil drop is injected at the bottom of the column using a plastic syringe. The images are recorded using a fast speed camera (200 fps), and the set-up is backlit using a LED panel. Behind the drop, we observe formation of droplets.

(a)

Fluid	Density (kg/L)	Property	Role
Corn Oil	0.92	Newtonian η =0.06 Pa.s^{-1}	Liquid drop
Polyacrylamide solution	1.13	Shear-thinning Viscoelastic	Surrounding Fluid

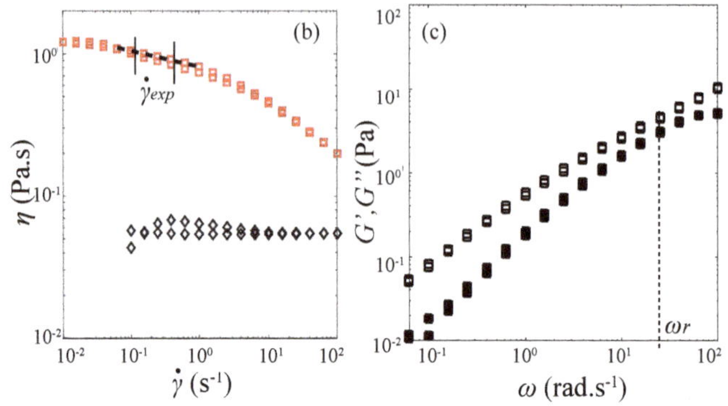

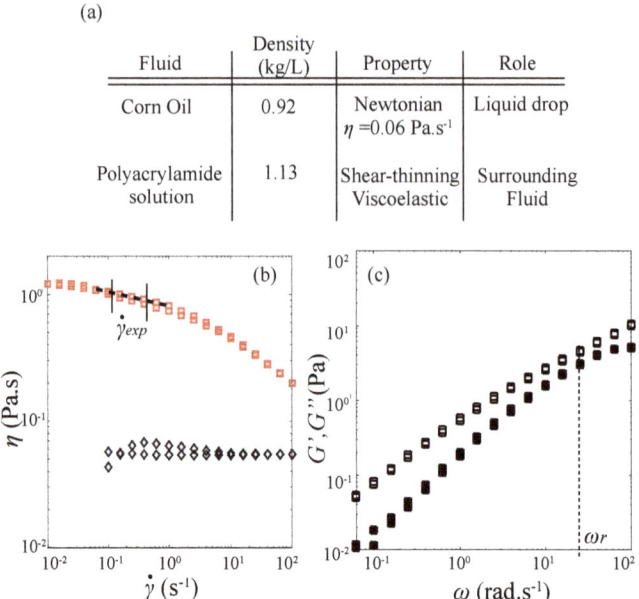

Figure 2. (**a**) table compiling the properties of the two fluids used. Since the oil drop has a lower density and a Newtonian behaviour, it will rise in the surrounding fluid consisting of a water-polyacrylamide solution, which is denser and has shear-thinning and viscoelastic properties; (**b**) measured viscosity η as a function of the shear-rate $\dot{\gamma}$ for the two fluids used. The diamonds represent the oil drop (Newtonian), the squares the polyacrylamide solution (Shear-thinning). $\dot{\gamma}_{exp}$ is the shear-rate range in our experiment and the dashed line corresponds to the power-law fit; (**c**) elastic modulus G' (full squares) and viscous modulus G'' (empty squares) as a function of the oscillation frequency for the polyacrylamide solution. A viscoelastic behaviour is clearly observed. The relaxation frequency ω_r, is estimated when the two moduli are the closest. In (**b**,**c**), the measurements are performed both increasing and decreasing the shear-rate/oscillation frequency.

3. Experimental Observations: Different Regimes

The experiment is performed by injecting oil at the bottom of the fluid column, using a plastic syringe. The oil volume V is not measured a priori, instead it is estimated by image analysis, considering that, for small drops, the volume corresponds to the one of a sphere of diameter D:$V = \pi D^3/6$ (Figure 3a), and, for the bigger ones, it is the sum of a cone of height H and a hemisphere of diameter D:$V = \pi D^2 H/12 + \pi D^3/12$ (Figure 3b–e). To ensure good statistics, we reproduce the experiment 50 times varying the volume V from 0.01 to 0.47 mL. We detect the position of the front of the drop to determine its velocity. Figure 4a presents the evolution of the vertical position z of the front of the drop as a function of time t for a drop of volume $V = 0.36$ mL. For all drops, we observe that the vertical position is linear in time t; the rising velocity U is computed by a simple linear regression. Figure 4b shows the rising velocity U of the drop as a function of its volume V. The rising velocity increases slowly with the volume until it reaches a critical volume ($V_c = 0.13$ mL). At this volume, a small velocity jump is observed, which has already been reported in literature as the velocity discontinuity [14–18,22]. This appears for bubbles and drops rising in a viscoelastic surrounding fluid, and is directly linked with the appearance of a negative wake behind the bubble/drop. Above the critical volume V_c, the rising velocity U increases more rapidly with the volume V. Considering the non-Newtonian properties of the surrounding fluid (shear-thinning and viscoelastic), it is not possible to predict the shape of the curve over this critical volume, but many other experimental examples have reported similar behaviour for drops or bubbles [14,20,21,23].

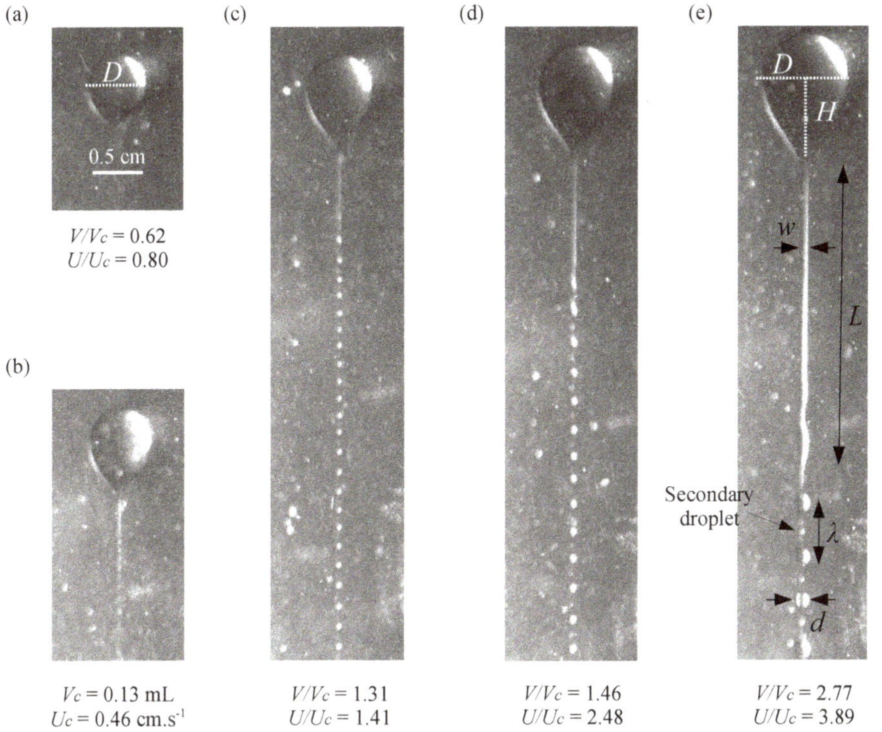

Figure 3. Different regimes observed: (**a**) before the tail appears; (**b**) at the critical volume $V_c = 0.13$ mL, where the tail appearance is. We can see very small droplets appearing behind the tail of the main drop; (**c**–**e**) instability for different volumes. We can observe that the tail length L, the width of the tail w, the distance between two droplets λ and the droplets diameter d increases with the volume. Those are defined in (**e**) and this will be discussed in detail in Section 4.

In terms of dimensionless numbers, it is common to use the Reynolds number Re and the Deborah number De. The Reynolds number compares the inertial forces of the flow with the viscous ones: $\mathrm{Re} = \rho UD/\eta = \rho UD/K\dot{\gamma}^{n-1}$, where ρ is the density of the surrounding fluid, U the velocity of the drop, D the diameter of the drop and η the viscosity. In our case, the fluid is shear thinning, so the viscosity changes with the shear rate. A common way to account for this problem is to define the shear-rate as the ratio of velocity and diameter of the drop $\dot{\gamma} = U/D$, and to use this in the rheological measurements using the formula $\eta = K\dot{\gamma}^{n-1}$ where $n = 0.87$ and $K = 1.10$ (see Section 2). We obtain a modified Reynolds number scaling as $\mathrm{Re} = U^{2-n}\rho D^n/K$ (see, for instance, [33]). This gives us a Reynolds number varying from 5×10^{-3} to 2.39. This small Reynolds number shows that inertial effects are small. The Deborah number compares the viscoelastic relaxation time and the observation time scale: $\mathrm{De} = \tau_r/\tau_o$. We can define the observation time scale as the inverse of the shear rate: $\tau_o = 1/\dot{\gamma} = D/U$ and the relaxation time is defined in Section 2: $\tau_r = 2\pi/\omega_r = 0.25$ s. In our experiment, we have the Deborah number varying from 5.6×10^{-2} for the largest drops to 10.6×10^{-2} for the smallest ones. This range of values indicates that elastics effects are small but not negligible. Note that experiments were conducted only with one container size. Since the Reynolds number is small, the walls will affect the terminal velocity of the drops. Considering the Faxen series correction (see Mendoza-Fuentes et al., [34]), the terminal velocity will be smaller by 51% for the case of the largest drop. Note that the correction is valid only for spherical particles in Newtonian fluids; hence, we expect the wall effects to be present but to be smaller than this value. Figure 3 shows the different regimes of the drops as the volumes V increases. First, at small volumes (Figure 3a, the scale is reported on this image and is the same for all), the drop is spherical and no significant shape alterations are detected. When the drop reaches the the critical volume V_c (Figure 3b), a tail appears. According to Ortiz et al. [20] and Zenit and Feng [17], the appearance of the tail coincides with the formation of a negative wake. This tail will undergo a capillary instability where droplets are produced. At the critical volume, the tail is very small as are the droplets released. For the rest of the article, we will use the term droplets for the liquid released behind the tail and drop for the main one. When the volume is increased (Figure 3c–e), we observe that the tail grows bigger in length L and width w, as well as the droplets diameter d and the distance between two droplets λ. Those values are defined in Figure 3e. One important fact to note is that the volume of the original drop V is not constant since it releases droplets. This will be further discussed in Section 5. Note that secondary smaller droplets appear for the biggest drops ($V/V_c > 1.9$, Figure 3e). The formation of such secondary droplets has been discussed previously for the Newtonian case in [24]. In this article, we will focus only on the main droplets' formation.

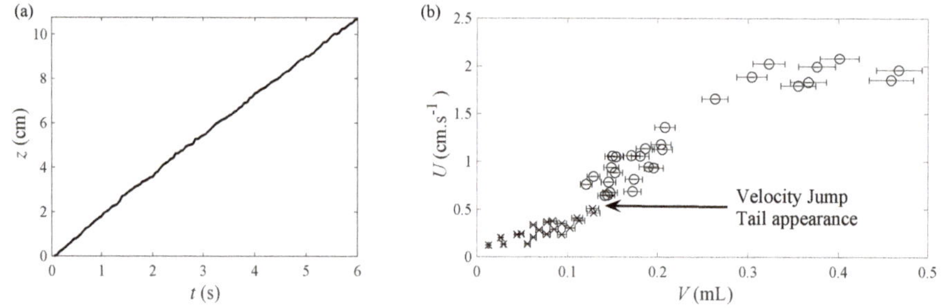

Figure 4. (**a**) position of the front of the drop as a function of time. This example corresponds to Figure 3e. We observe that the rising velocity U stays constant over the 12 cm experiment height; (**b**) rising velocity U of the drops, as a function of the volume V. We observe a small velocity jump at the moment of the tail appearance for a critical volume $V_c = 0.13$ mL and a critical velocity $U_c = 0.46$ cm·s^{-1}.

4. Droplet Formation

4.1. Tail Size

From the images, the size of the tail behind the drop can be readily measured. We can obtain its width w and its length L, as defined in Figure 3e.

Since we photograph the drop in a terminal condition, we do not observe the initial formation of the tail. As for the volume of the drop, the length L and width w of the tail might change during the rising, since droplets are emitted, but, once again, we did not observe a significant reduction of either the length or the width of the tail over the height of the camera window (12 cm). We measured the length L and the width w once the tail was fully visible in the images. Figure 5a shows the length of the tail as a function of the velocity of the drop U. We observe a breakup of the end of the tail leading to droplets' emission at a distance going from 0.5 to 3.6 cm from the main drop. This distance (which corresponds to what we called the tail length L) will vary linearly with the velocity U, with a slope of 1.5 s (dashed line, Figure 5a). This shows that the length of the tail depends directly on the velocity of the drop, via the negative wake. The slope corresponds to the time needed for the drop to move a distance L.

Figure 5b shows the tail aspect ratio L/w as a function of the drop velocity U. This aspect ratio is between 10 (for the drop at the transition), up to 50 for the largest drops. This shows that the tail is very long compared to its width, the width being for the smallest tail of about 0.035 cm to 0.075 cm for the biggest one. The fact that the curve increases shows that the tail will grow more rapidly with the velocity U in length L than in width w. We can explain this by the fact that the main driving force for the length of the tail is the strong non-Newtonian behaviour of the fluid, while the width of the tail is limited by the capillary length.

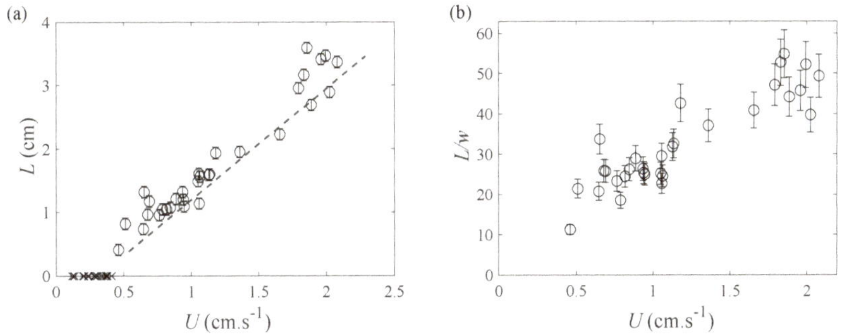

Figure 5. (**a**) tail length L as a function of the velocity of the drop U. We see clearly a critical velocity U_c where the tails appears (corresponding to a critical volume V_c). The tail length grows linearly with the speed of the drop (dashed line); (**b**) tail aspect ratio L/w as a function of the drop velocity V. The tail has a very elongated shape.

4.2. Emission Period and Wavelength

Figure 6a shows the average time Δt_e (or emission period) between two droplets emitted, as a function of the velocity of the drop. Except at the critical velocity V_c, the period of emission of droplets Δt_e is roughly constant and has a value contained between 0.2 and 0.3 s. This results from a competition between the width w of the tail and the velocity U of the drop. For small velocity, the tail is thinner, so it would tend to break more easily, and, at higher velocity, the drainage of the tail is more rapid, which also helps the breakup. At the end, the time between the emission of two droplets emitted will be roughly the same for all velocity.

Figure 6b shows the average distance between two droplets λ, as a function of the velocity of the drops U. This distance increases importantly with velocity, which is in agreement with the constant emission period Δt_e: since the tail velocity is the same as the drop velocity U (stationary regime), and the emission time between two drops is almost constant; this implies that the distance between two drops will increase with the velocity of the drop. We have a linear relation between λ and U for the drops over the critical volume V_c, with a slope of $\Delta t_e \approx 0.27$ s (dashed line, Figure 6b), which is in accordance with Figure 6a. We can compute the frequency of emission of the droplets, if we assume that the velocity is constant during one run, which gives us $f_e = 1/\Delta t_e \approx 3.7\ \mathrm{s}^{-1}$.

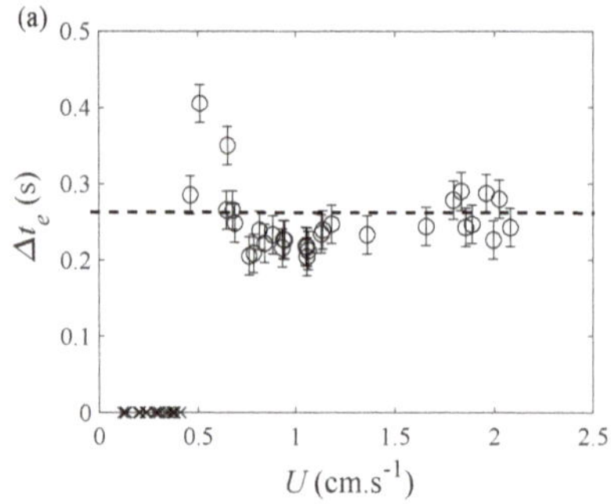

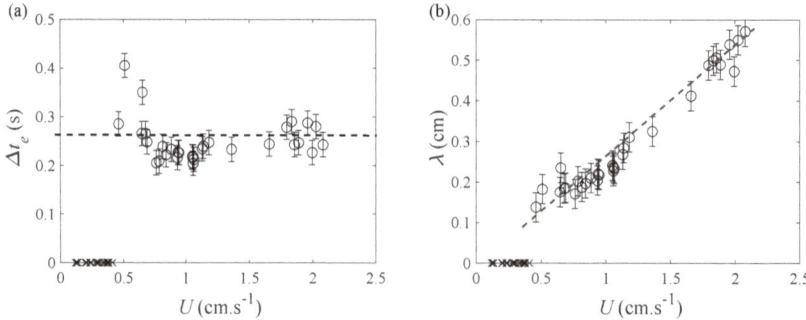

Figure 6. (**a**) period of emission Δt_e of the droplets (average time between two droplets appearance), as a function of the drop velocity U. Except close to the critical volume, this period seems roughly constant (dashed line, $\Delta t_e = 0.27$ s), which corresponds at a frequency of emission f_e of 3.7 Hz; (**b**) wavelength λ (average distance between two droplets), as a function of the drop velocity U. The dashed line represents the linear adjustment.

4.3. Droplet Size

We also analyzed the size of the droplets created behind the main drop. Once again, we observe that the size of the droplets is constant over the course of one experiment (for one given drop volume V). The contrast being better for the droplets than the drop, the uncertainty on the diameter gets to about one pixel, but it is still important in comparison with the droplet diameter. To reduce the error, we measure the diameter for 10 different droplets, which decrease significantly the error, down to an estimated 5 percent. The error bars have been reported on Figure 7. We used the same method to estimate the width of the tail, measuring the width at different heights and then averaging. Figure 7a shows the average volume of the droplets V_d as a function of the volume of the main drop V. The volume of the droplets V_d has been computed assuming that the droplet is a sphere: $V_d = \pi d^3/6$, where d is the diameter of the droplets (see Figure 3e). We observe that the volume of the droplets increases linearly with the volume of the drop. This will be used to calculate the volume loss of the main drop over time (Section 5.2).

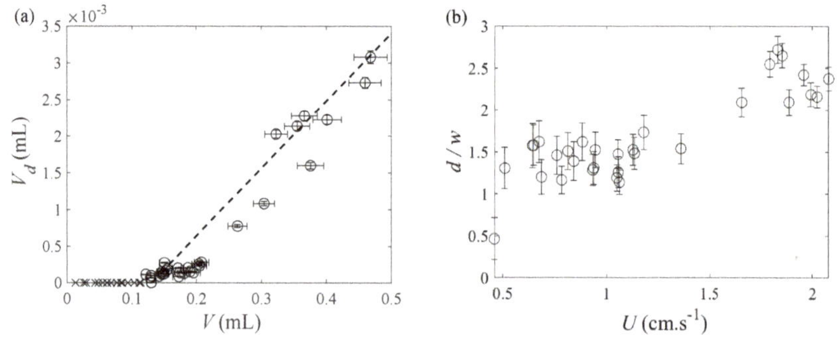

Figure 7. (**a**) mean volume of the droplets V_d as a function of the volume of the drop V. We observe an important increase which is coherent with Figure 3. The dashed line represents the linear regression, which will be used in Section 5; (**b**) droplet diameter d divided by the tail width w as a function of the drop velocity U. We observe that, except for the critical case, the diameter of the droplets is always bigger than the tail width.

Figure 7b shows the normalized droplet diameter d/w as a function of the velocity of the drop U. We observe that this ratio increases with the velocity, and, most of all, its value is always larger than one (except for one point at the transition), which means that the droplets are wider than the tail before it breaks. This can be explained by a simple mass conservation argument. The volume of oil before the break corresponds to the volume of a column of width w and height λ (for one wavelength), and also to the volume of one droplet of diameter d. We can write this volume as $V_d = \pi/4\lambda w^2 = \pi d^3/6$. Therefore,

$$\frac{d^2}{w^2} = 3/2\frac{\lambda}{d}. \tag{1}$$

Considering that the distance of between two droplets is much bigger than the size of the droplet except at the critical volume (see Figure 3), we have $\lambda/d > 1$, and so the diameter of the droplet d will be bigger than the width of the tail w.

5. Discussion

5.1. Tail Appearance and Breakup

We observe that a tail appears behind the drop for a volume larger than the critical one, $V_c = 0.13$ mL. This critical volume corresponds to the appearance of a negative wake behind the drop. This negative wake has been already studied in various cases, for bubbles [14–16,18,20,21,23], and for drops [11,17,20]. The main difference between the drop and the bubble case is that the interfacial tension between the air and the liquid is larger than that for two liquids. The bubble shape will then remain the same over the course of the experiment, while in the case of a drop, this interface is more deformable, and the tail will be able to grow due to the negative wake [20]. This negative wake will be more important as the velocity U increases, and the width of the tail w will also increase. The tail will then grow to a length L, where it will break up into small droplets. The size of these droplets is controlled by their emission frequency f_e, or its corresponding wave length, λ.

This breakup is similar to the Rayleigh–Plateau instability which arises from capillary effects [24,25]. The temporal evolution of this rupture, which fixes the length of the tail L, is hard to predict, as it takes into account the velocity of the fluid inside the tail. We do not have access to this velocity (it would require Particle Image Velocimetry (PIV) in the oil phase), and it is hard to predict it since it results from the negative wake. The fact that the length L increases with the velocity U is not trivial, but we can assume that the breakup occurs when the oil at the tip of the tail has a zero velocity. This would be in agreement with the negative wake increasing with the velocity of the drop. One way to describe such type of instabilities is to look at the capillary number and capillary length. The capillary number is the ratio between the capillary and the viscous forces: $\text{Ca} = \eta U'/\sigma$, where σ is the interfacial tension between the two fluids. In this case, U' is the speed of the droplets and not the velocity of the drop, but since we are in a stationary regime, we have $U' = U$. Since we have a proportionality between the wavelength λ, and the velocity $U = U'$, and the shear-thinning behaviour is small, we have a direct proportionality between the capillary number Ca and the wavelength λ:

$$\lambda = \Delta t_e U' = k \cdot \text{Ca}, \tag{2}$$

where $k = \Delta t_e \sigma/\eta$ is a constant coefficient with the dimension of a length. An important difficulty is to determine the interfacial tension σ between the two phases. In the literature, the corn oil surface tension is reported to be $\sigma_{oil} \approx 33.5$ mN/m [35] and for the polyacrylamide solution $\sigma_{PAAM} \approx 75$ mN/m [36]. At the first order, Antonoff's rule gives the interfacial tension between the two phases $\sigma = |\sigma_{PAAM} - \sigma_{oil}| \approx 41.5$ mN/m. By considering that $\eta \approx 0.9$ Pa.s, we obtain that $k \approx 1.24$ cm. The capillary length can be defined as $l_c = \sqrt{\sigma/\Delta\rho g} \approx 1.29$ cm, where $\Delta\rho$ is the density difference between the two fluids, and g is the gravitational acceleration. Therefore, $k \approx l_c$, which is consistent with the assumption that, indeed, the breakup of the tail results from capillary instability. This scaling relies on important assumptions, notably on the value of the interfacial tension. In addition, it does not take into account the viscoelastic properties of the surrounding fluid. Nevertheless, these scaling arguments indicate that indeed the tail is fragmenting mainly as a result of capillary instability and the viscoelastic effects are secondary.

5.2. Volume Loss

It is important to evaluate the role of volume change for the main drop, resulting from the droplets emitted at the tail. In all cases, we assumed that the drop volume V was constant. This assumption is supported by two facts. First, in Figure 4a, the bubble rises at a constant velocity, which would not have been the case if the volume had varied significantly. Secondly, Figure 7a show that the volume of the droplets V_d remains smaller than 0.65 percent of the main drop volume, for the largest drops. In this case, only 20 droplets are emitted over the experiment, which makes (in the worst case scenario) a volume loss around 13 percent of the initial drop volume.

A simple model is proposed to predict the volume of the droplets in an infinitely long liquid column. First, by using the linear regression in Figure 7a, we can predict the volume of a droplet knowing the volume of the main drop. Then, assuming that the emission frequency of droplets is constant $f_e = 3.7\ \mathrm{s}^{-1}$ (Figure 6a), we can write the following differential equation for the volume change:

$$\frac{\mathrm{d}V}{\mathrm{d}t} = -f_e V_d = -f_e(\alpha \cdot V + \beta), \tag{3}$$

where $\alpha = 9.2 \times 10^{-3}$ and $\beta = -1.2 \times 10^{-3}$ mL are the slope and intercept of the dashed line in Figure 7a. Integrating, we obtain

$$V(t) = V_0 \exp(-f_e \cdot \alpha \cdot t) + \frac{\beta}{\alpha}(\exp(-f_e \cdot \alpha \cdot t) - 1), \tag{4}$$

where V_0 is the initial drop volume (taken to be 0.5 mL). This expression, if used carelessly, will predict a negative volume value for long times; however, one must consider that the droplets will no longer be emitted once the volume $V(t)$ reaches the critical volume ($V_c = 0.13$ mL). The drop will then rise with a constant volume V_c and a constant velocity $U_c \approx 0.46\ \mathrm{cm{\cdot}s^{-1}}$. Figure 8a shows the volume evolution $V(t)$ as a function of time t. The critical volume V_c is reached at a time $t_c = 26$ s.

We can use a linear regression between the volume and the velocity over the critical volume in Figure 4b which gives $U(t) = 7.6V(t) - 0.52$. Figure 8b shows the velocity of the drop U, as a function of time t, the velocity decreases from $3.3\ \mathrm{cm{\cdot}s^{-1}}$ to $U_c = 0.46\ \mathrm{cm{\cdot}s^{-1}}$. This is clearly only a first order approximation, since the relation between the volume and velocity is most likely nonlinear. It allows us to continue the integration. We can then compute the position $z(t)$ of the drop as a function of time as:

$$z(t) = \int_0^t U \mathrm{d}t' = \int_0^t (7.6V(t) - 0.52)\mathrm{d}t'. \tag{5}$$

For simplicity, we will not write down this integral (it implies exponential integrals). Figure 8c shows the position of the droplets emitted z_d, as a function of time t. The droplets are emitted every $\Delta t_e = 0.27$ s, and the marker size is proportional to the volume of the droplets. The drop will reach its critical volume at a position $z_c = 43.4$ cm, and then will rise at its constant velocity, without emitting new droplets. The volume of the droplets emitted will vary from $V_d = 0.0034$ mL at the beginning, and will tend to 0 when we approach the critical volume. We must emphasize that the calculation above is only valid for the two-fluid combination considered here. However, the same general behaviour is expected for a Newtonian/non-Newtonian combination. Clearly, more experiments are needed to extend the parametric range of validity. This simple model gives us an order of magnitude of what should be expected in terms of time t_c and height z_c for the bubble to reach its critical volume V_c. This is in agreement with what was shown before: t_c is much bigger than the time of our experiment (10 s), and z_c is also much bigger than the 12 cm where we observed the rise, so the model holds some consistency. This simple model could lead to some applications. For instance, we could imagine a device where one would like to encapsulate oil droplets (containing another substance to analyze or to use as a reactant for example) with droplets varying in size. With this simple two-fluid configuration, and by choosing well both fluids (which would require more experiments and more general understanding), one could construct such device, which would be very easy to use since the only input would be the volume of the main drop. Since the emitted droplets would be small, their rising velocity would be small too. Hence, their capture would be relatively simple. Additionally, one could use a surrounding yield-stress viscoelastic fluid, so that the encapsulated droplets would be completely trapped and then easily manipulated.

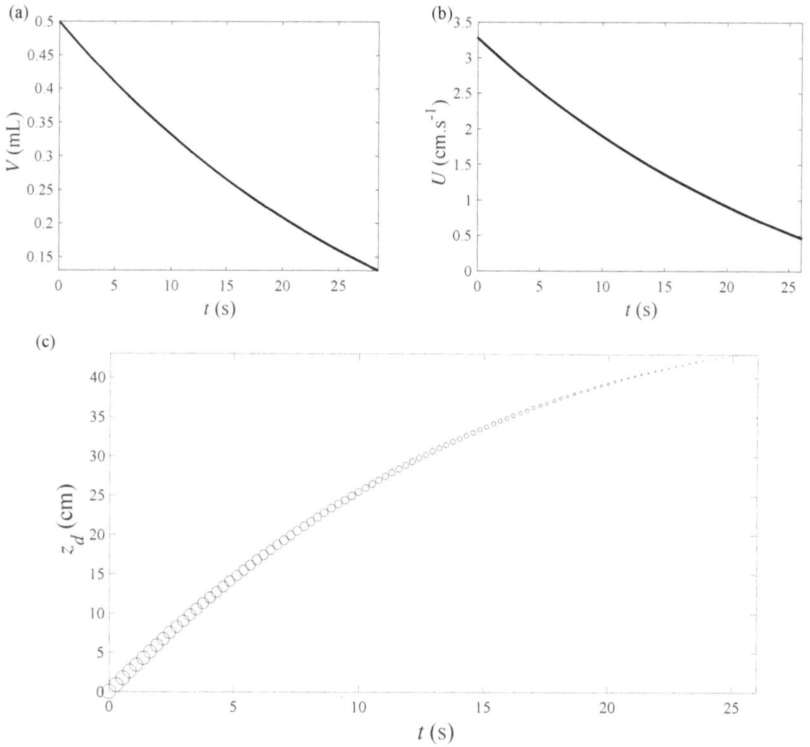

Figure 8. (**a**) evolution of the volume V as a function of time t. The critical volume $V_c = 0.13$ mL is reached after a time $t_c = 28.5$ s; (**b**) velocity of the droplet U as a function of time t; (**c**) velocity of the droplet U as a function of time t; (**c**) position of the droplets z_d as a function of time t. The size of the markers represents the volume of each droplet.

6. Conclusions and Perspectives

In this article, we investigated the instability occurring at the tail of a Newtonian drop rising in a viscoelastic fluid. We observe that this leads to the formation of small droplets, which are controlled in size by the velocity of the drop, which in turn depends on its volume. It is interesting to note that such behaviour (cusp and tail formation with fragmentation) had not been discussed previously in this combined manner. In this article, we use the results of a very particular case where the rising drop presents a velocity discontinuity and a negative wake (corn oil in a polyacrylamide solution), so the physical properties of the fluids are fixed. We described the appearance of a tail resulting from the negative wake over a critical volume and its breakup due to capillary forces. This limits our results to the ranges of Reynolds and Deborah number we used. To provide a more general understanding, more experiments are needed, by changing both fluids. A number of open questions remain, notably on the role of the viscosity ratio between both fluids, the interfacial tension, the quantitative role of the surrounding fluid elasticity, for both the tail formation and the breakup, as well as the role of elongational rheology (since we are dealing with polymer solutions). An exhaustive study would require a very important number of experiments, since changing fluids will influence all the properties at once (density, viscosity, elasticity, critical volume, etc.). We plan to pursue such experiments in the future. Finally, we proposed a simple model, based on the volume of the emitted droplets, that could have some application to encapsulate droplets with varying size, with just one input: the volume of the main drop.

Author Contributions: Investigation, R.P.; Writing—Original Draft Preparation, R.P.; Supervision, R.Z.

Funding: Raphael Poryles acknowledges the support of DGAPA-UNAM for postdoctoral support.

Acknowledgments: The rheological measurement were performed by Elsa De la Calleja from Instituto de Investigaciones en Materiales, Universidad Nacional Autónoma de México.

Conflicts of Interest: The authors declare no conflict of interest.

References

1. Wu, L.; Chen, P.; Dong, Y.; Feng, X.; Liu, B.F. Encapsulation of single cells on a microfluidic device integrating droplet generation with fluorescence-activated droplet sorting. *Biomed. Microdevices* **2013**, *15*, 553–560. [CrossRef] [PubMed]
2. Teh, S.Y.; Lin, R.; Hung, L.H.; Lee, A.P. Droplet microfluidics. *Lab Chip* **2008**, *8*, 198–220. [CrossRef] [PubMed]
3. Theberge, A.B.; Courtois, F.; Schaerli, Y.; Fischlechner, M.; Abell, C.; Hollfelder, F.; Huck Wilhelm, T.S. Microdroplets in Microfluidics: An Evolving Platform for Discoveries in Chemistry and Biology. *Angew. Chem. Int. Ed.* **2013**, *49*, 5846–5868. [CrossRef] [PubMed]
4. Mark, D.; Haeberle, S.; Roth, G.; Von Stetten, F.; Zengerle, R.L. Microfluidic Lab-on-a-Chip Platforms: Requirements, Characteristics and Applications. In *Microfluidics Based Microsystems;* Kakaç, S., Kosoy, B., Li, D., Pramuanjaroenkij, A., Eds.; Springer: Dordrecht, The Netherlands, 2010; pp. 305–376, ISBN 978-90-481-9029-4.
5. Garstecki, P.; Fuerstman, M.J.; Stone, H.A.; Whitesides, G.M. Formation of droplets and bubbles in a microfluidic T-junction—Scaling and mechanism of break-up. *Lab Chip* **2006**, *6*, 437–446. [CrossRef] [PubMed]
6. De Menech, M.; Garstecki, P.; Jousse, F.; Stone, H. Transition from squeezing to dripping in a microfluidic T-shaped junction. *J. Fluid Mech.* **2008**, *595*, 141–161. [CrossRef]
7. Arigo, M.T.; McKinley, G.H. An experimental investigation of negative wakes behind spheres settling in a shear-thinning viscoelastic fluid. *Rheol. Acta* **1998**, *37*, 307–327. [CrossRef]
8. Bisgaard, C.; Hassager, O. An experimental investigation of velocity fields around spheres and bubbles moving in non-Newtonian liquids. *Rheol. Acta* **1982**, *21*, 537–539. [CrossRef]
9. Broadbent, J.; Mena, B. Slow flow of an elastico-viscous fluid past cylinders and spheres. *Chem. Eng. J.* **1974**, *8*, 11–19. [CrossRef]
10. Caswell, B.; Manero, O.; Mena, B. Recent developments on the slow viscoelastic flow past spheres and bubbles. *Rheol. Rev* **2004**, 197–223.
11. Chhabra, R.P. *Bubbles, Drops and Particles in Non-Newtonian Fluids;* CRC Press: Boca Raton, FL, USA, 1993.
12. Manero, O.; Mena, B. On the slow flow of viscoelastic fluids past a circular cylinder. *J. Non-Newton. Fluid Mech.* **1981**, *9*, 379–387. [CrossRef]
13. Mena, B.; Manero, O.; Leal, L.G. The influence of rheological properties on the slow flow past spheres. *J. Non-Newton. Fluid Mech.* **1987**, *26*, 247–275. [CrossRef]
14. Herrera-Velarde, J.R.; Zenit, R.; Chehata, D.; Mena, B. The flow of non-Newtonian fluids around bubbles and its connection to the jump discontinuity. *J. Non-Newton. Fluid Mech.* **2003**, *111*, 199–209. [CrossRef]
15. Rodrigue, D.; De Kee, D. Bubble velocity jump discontinuity in polyacrylamide solutions: A photographic study. *Rheol. Acta* **1998**, *37*, 307–327. [CrossRef]
16. Rodrigue, D.; De Kee, D.; Chan Man Fong, C. Bubble velocities: further developments on the jump discontinuity. *J. Non-Newton. Fluid Mech.* **1998**, *79*, 45–55. [CrossRef]

17. Zenit, R.; Feng, J.J. Hydrodynamic Interactions Among Bubbles, Drops, and Particles in Non-Newtonian Liquids. *Annu. Rev. Fluid Mech.* **2018**, *50*, 505–534. [CrossRef]
18. Astarita, G.; Apuzzo, G. Motion of gas bubbles in non-Newtonian liquids. *AIChE J.* **1965**, *11*, 815–820. [CrossRef]
19. Pilz, C.; Brenn, G. On the critical bubble volume at the rise velocity jump discontinuity in viscoelastic liquids. *J. Non-Newton. Fluid Mech.* **2007**, *145*, 124–138. [CrossRef]
20. Ortiz, S.L.; Lee, J.S.; Figueroa-Espinoza, B.; Mena, B. An experimental note on the deformation and breakup of viscoelastic droplets rising in non-Newtonian fluids. *Rheol. Acta* **2016**, *55*, 879–887. [CrossRef]
21. Soto, E.; Goujon, C.; Zenit, R.; Manero, O. A study of velocity discontinuity for single air bubbles rising in an associative polymer. *Phys. Fluids* **2006**, *18*, 121510. [CrossRef]
22. Fraggedakis, D.; Pavlidis, M.; Dimakopoulos, Y.; Tsamopoulos, J. On the velocity discontinuity at a critical volume of a bubble rising in a viscoelastic fluid. *J. Fluid Mech.* **2016**, *789*, 310–346. [CrossRef]
23. Hassager, O. Negative wake behind bubbles in non-Newtonian liquids. *Nature* **1979**, *279*, 402–403. [CrossRef] [PubMed]
24. Lister, J.R.; Stone, H.A. Capillary breakup of a viscous thread surrounded by another viscous fluid. *Phys. Fluids* **1998**, *10*, 2758–2764. [CrossRef]
25. De Gennes, P.G.; Brochard-Wyart, F.; Quéré, D. *Capillary and Wetting Phenomena—Drops, Bubbles, Pearls, Waves*; Alex, R., Trans.; Springer: Berlin/Heidelberg, Germany, 2002.
26. Deblais, A.; Velikov, K.P.; Bonn, D. Pearling instabilities of a viscoelastic thread. *Phys. Rev. Lett.* **2018**, *120*, 194501. [CrossRef] [PubMed]
27. Clasen, C.; Eggers, J.; Fontelos, M.A.; Li, J.; McKinley, G. The beads-on-string structure of viscoelastic threads. *J. Fluid Mech.* **2006**, *556*, 283–308 [CrossRef]
28. Skelland, A.H.P.; Raval, V.K. Drop size in power law non-Newtonian systems. *Can. J. Chem. Eng.* **1972**, *50*, 41–44. [CrossRef]
29. Kitamura, Y.; Takahashi, T. Breakup of jets in power law non-Newtonian–Newtonian liquid systems. *Can. J. Chem. Eng.* **1982**, *60*, 732–737. [CrossRef]
30. Teng, H.; Kinoshita, C.M.; Masutani, S.M. Prediction of droplet size from the breakup of cylindrical liquid jets. *Int. J. Multiph. Flow* **1995**, *21*, 129–136. [CrossRef]
31. Barnes, H.A.; Hutton, J.F.; Walters, K. *An Introduction to Rheology*; Elsevier: New York, NY, USA, 1989.
32. Ghannam, M.T.; Esmail, M.N. Rheological properties of aqueous polyacrylamide solutions. *J. Appl. Polym. Sci.* **1998**, *69*, 1587–1597. [CrossRef]
33. Palacios-Morales, C.; Zenit, R. The formation of vortex rings in shear-thinning liquids. *J. Non-Newton. Fluid Mech.* **2013**, *194*, 1–13. [CrossRef]
34. Mendoza-Fuentes, A.J.; Manero, O.; Zenit, R. Evaluation of drag correction factor for spheres settling in associative polymers. *Rheol. Acta* **2010**, *49*, 979–984. [CrossRef]
35. Esteban, B.; Riba J.-R.; Baquero, G.; Puig, R.; Rius, A. Characterization of the surface tension of vegetable oils to be used as fuel in diesel engines. *Fuel* **2012**, *102*, 231–238. [CrossRef]
36. Hu, R.Y.Z.; Wang, A.T.A.; Hartnett, J.P. Surface tension measurement of aqueous polymer solutions. *Exp. Therm. Fluid Sci.* **1991**, *4*, 723–729. [CrossRef]

fluids

MDPI

Article

A Numerical Study of Particle Migration and Sedimentation in Viscoelastic Couette Flow

Michelle M. A. Spanjaards [1], Nick O. Jaensson [2], Martien A. Hulsen [1] and Patrick D. Anderson [1,*]

[1] Department of Mechanical Engineering, Eindhoven University of Technology, P.O. Box 513, 5600 MB Eindhoven, The Netherlands; m.m.a.spanjaards@tue.nl (M.M.A.S.); m.a.hulsen@tue.nl (M.A.H.)

[2] Department of Materials, ETH Zurich, Vladimir-Prelog-Weg 5, 8093 Zurich, Switzerland; nick.jaensson@mat.ethz.ch

* Correspondence: p.d.anderson@tue.nl

Received: 16 January 2019; Accepted: 2 February 2019; Published: 11 February 2019

Abstract: In this work, a systematic investigation of the migration of sedimenting particles in a viscoelastic Couette flow is presented, using finite element 3D simulations. To this end, a novel computational approach is presented, which allows us to simulate a periodic configuration of rigid spherical particles accurately and efficiently. To study the different contributions to the particle migration, we first investigate the migration of particles sedimenting near the inner wall, without an externally-imposed Couette flow, followed by the migration of non-sedimenting particles in an externally-imposed Couette flow. Then, both flows are combined, i.e., sedimenting particles with an externally-imposed Couette flow, which was found to increase the migration velocity significantly, yielding migration velocities that are higher than the sum of the combined flows. It was also found that the trace of the conformation tensor becomes asymmetric with respect to the particle center when the particle is initially placed close to the inner cylinder. We conclude by investigating the sedimentation velocity with an imposed orthogonal shear flow. It is found that the sedimentation velocity can be both higher or lower then the Newtonian case, depending on the rheology of the suspending fluid. Specifically, a shear-thinning viscosity is shown to play an important role, which is in-line with previously-published results.

Keywords: migration; sedimentation; viscoelasticity

1. Introduction

Suspensions of particles in viscoelastic fluids are encountered in many natural, biological, and industrial processes. Examples of these suspensions are blood, paint, and pharmaceuticals. Therefore, the behavior of particles suspended in a single viscoelastic fluid is of great interest. Many practical applications also show the occurrence of cross-streamline migration of particles. Examples are separation of particles with different sizes, suspensions in microfluidic devices where migration can lead to non-homogenous particle distributions, cells in blood causing clots, which are obstructing the flow, and other biosensing applications [1]. Furthermore, Couette flow devices are often used as a rheometer, to measure the rheology of suspensions. Therefore, the migration behavior of the particles is of great interest, since migration of particles will change the rheology of the fluid.

Understanding the sedimentation behavior of particles in a viscoelastic shear flow can for example be used to optimize the hydraulic fracturing process for obtaining natural gas. In this process, solid particles in polymeric solutions are pumped into hydraulically-induced fractures under high pressure. The particles must keep the fractures open in order to increase the rate of gas recovery. In vertically-oriented fractures, sedimentation causes the particle to settle. Therefore, sedimentation should be limited in order to limit settling and place the particles far into the fracture to increase gas recovery [2].

Migration of a sedimenting particle in a channel near a vertical wall was investigated by Singh and Joseph [3]. They found that a falling particle initially placed close enough to the wall will migrate towards the wall. However, in a Newtonian liquid, the sphere moves away from the vertical wall and attains a steady position between the channel center and the wall.

When the particle was placed in a shear flow, D'Avino et al. [4] found that three different regimes can be recognized in the migration of the particle, depending on the distance of the particle from the wall, where the migration velocity is initially linear with the particle position, but increases when the particle approaches the wall and decreases again when the particle is in very close proximity to the wall.

Migration of a particle in a viscoelastic fluid in a Couette flow was numerically and experimentally investigated by D'Avino et al. [5]. It was found that elasticity promotes migration of the particle towards the outer cylinder for almost every initial position. Only if the particle was initially placed very close to the inner cylinder, it would move towards it. Furthermore, a maximum in the migration velocity was found when the particle moves closer to the outer cylinder. High Newtonian solvent viscosity and shear thinning reduce the migration rate.

The sedimentation of particles in a flow without shear has been studied extensively. A dimensionless number called the sedimentation Weissenberg number relates the polymer relaxation time scale to the flow time scale. It has been shown that at high values for Λ, the deformation of the polymer is dominated by the flow due to the movement of the particle. This can cause significant extension of the polymer around the particle, causing pronounced elastic effects in the wake of the sphere that decrease the particle mobility [6–8].

Many research works have also been devoted to the sedimentation of a particle in a shear flow [9]. Van den Brule et al. [10] performed experiments to gain more insight into the effects of fluid elasticity on the settling velocity of spherical particles suspended in viscoelastic fluids. Here, the settling velocity of a spherical particle was measured in three different fluids: a Newtonian fluid, a Boger fluid and a shear-thinning fluid. From this study, it could be concluded that fluid elasticity plays a significant role in reducing the settling velocity of a spherical particle in a viscoelastic fluid. Shear-thinning, however, increased the settling velocity of the particles, since now the viscosity decreases with increasing shear rate. This was also found in another study [11]. Padhy et al. [12] numerically investigated the effect of viscoelasticity on the settling velocity of a particle in a cross-shear flow. Padhy et al. [12] calculated the drag on a particle in their simulations and found that the drag in viscoelastic fluids is increased indirectly through the increase of viscous stresses. The dominant contribution to the viscous stresses becomes asymmetric when sedimentation and shear flow are combined, leading to a decrease in settling velocity. The coupling between the two flows needs to be non-linear for the increase in drag to be substantial. This non-linear coupling is only present in viscoelastic fluids.

Murch et al. [13] recently numerically investigated the growth of viscoelastic wings and how this reduces the particle mobility in a viscoelastic shear flow. They found that for high shear Weissenberg numbers and sedimentation Weissenberg numbers of $O(1)$, the particle mobility is decreased due to the growth of viscoelastic wings around the particle.

Migration is often a problem in flows that are more complex than simple shear, or pure sedimentation. In the current paper, a first step up in complexity is offered, by combining two flows that yield migration. The finite element method is used to investigate numerically the influence of both the sedimentation and shear Weissenberg number on the migration and sedimentation of a particle in a viscoelastic fluid under orthogonal shear in a Couette flow device. The mass of the particle is varied via the downward force on the particle to create different local shear rates around the falling particle, leading to different sedimentation Weissenberg numbers. Furthermore, the orthogonal shear rate is varied to create different shear Weissenberg numbers. The influence of both Weissenberg numbers on the sedimentation and migration of the particle is investigated, using a Giesekus model. Migration of the particle close to the inner cylinder of the Couette is studied systematically to see how the different flow types influence the migration of the particle and what happens if the two are combined.

2. Problem Statement

The shear flow is created in a Couette flow device, by rotating the outer cylinder with a velocity U_w, while keeping the inner cylinder stationary. When studying sedimentation, a force is applied to the particle in the z-direction. To minimize the computational costs, only a part of the Couette flow device is modeled, as is shown in Figure 1. Periodicity is assumed on surfaces S_2 and S_4. The top and bottom of the domain are assumed to be far enough away from the particle, to not feel the presence of the particle. The angle ϕ, together with the inner radius R_i, the outer radius R_o, and the height H define the size and shape of the domain. The spherical particle has radius a, and its boundary is denoted by ∂P. A cylindrical coordinate system will be used throughout this paper, with components, using the convention $[r, \theta, z]$. The computational domain of the Couette is denoted by Ω. Note that the positive z-direction is directed in the gravitation direction.

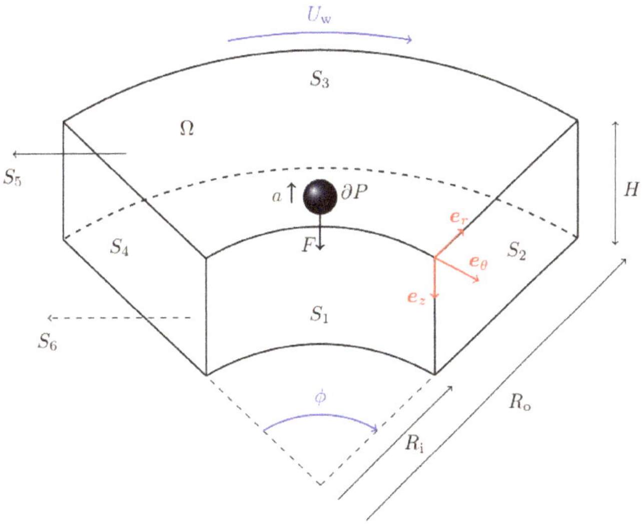

Figure 1. Problem description of a rigid, spherical particle suspended in a viscoelastic Couette flow. The Couette flow is set up by moving the outer cylinder in the tangential direction, with velocity U_w, while the inner cylinder remains stationary. Simultaneously, a force acts on the particle in the z-direction, causing it to sediment. The inner cylinder boundary is denoted by S_1, the outer cylinder boundary is denoted by S_3. The top and bottom of the Couette are denoted by S_5 and S_6, respectively. The periodic side boundaries are denoted by S_2 and S_4, where S_4 is rotated over an angle ϕ with respect to surface S_2.

2.1. Governing Equations

2.1.1. Balance Equations

It is assumed that the fluid is incompressible, that inertia can be neglected, and that there are no external body forces acting on the fluid. This leaves the following equations for the balance of momentum and balance of mass:

$$-\nabla \cdot \boldsymbol{\sigma} = \mathbf{0} \qquad \text{in } \Omega, \tag{1}$$

$$\nabla \cdot \boldsymbol{u} = 0 \qquad \text{in } \Omega, \tag{2}$$

where $\boldsymbol{u}$ is the fluid velocity and $\boldsymbol{\sigma}$ is the Cauchy stress tensor, which consists of three contributions:

$$\boldsymbol{\sigma} = -p\boldsymbol{I} + 2\eta_s\boldsymbol{D} + \boldsymbol{\tau}_p, \quad (3)$$

where p is the pressure, $\boldsymbol{I}$ the unit tensor, and η_s the solvent or Newtonian viscosity. Furthermore, $\boldsymbol{D} = (\nabla\boldsymbol{u} + \nabla\boldsymbol{u}^{\mathrm{T}})/2$ is the rate of deformation tensor, and $\boldsymbol{\tau}_p$ represents the viscoelastic stress tensor, given by:

$$\boldsymbol{\tau}_p = G(\boldsymbol{c} - \boldsymbol{I}), \quad (4)$$

where $\boldsymbol{c}$ is the conformation tensor and G the polymer modulus, which equals η_p/λ. Here, η_p is the polymer viscosity and λ the polymer relaxation time. The evolution of the conformation tensor $\boldsymbol{c}$ can be described by the following equation using the Giesekus model [14]:

$$\lambda\overset{\triangledown}{\boldsymbol{c}} + \boldsymbol{c} - \boldsymbol{I} + \alpha(\boldsymbol{c} - \boldsymbol{I})^2 = \boldsymbol{0}, \quad (5)$$

where α is the mobility parameter and $\overset{\triangledown}{()}$ denotes the upper-convected derivative, which is defined as:

$$\overset{\triangledown}{()} = \frac{D()}{Dt} - (\nabla\boldsymbol{u})^T \cdot () - () \cdot \nabla\boldsymbol{u}, \quad (6)$$

where $D()/Dt$ is the material derivative denoted by $D()/Dt = \partial()/\partial t + \boldsymbol{u} \cdot \nabla()$.

2.1.2. Particle Motion

It is assumed that the inertia of the particle can be neglected and that a force $\boldsymbol{F}_{\mathrm{ext}} = F\boldsymbol{e}_z$ is applied to the particle. The balance of forces and the balance of torques that act on the particle can be expressed as follows:

$$\int_{\partial P} \boldsymbol{\sigma} \cdot \boldsymbol{n}_p \, \mathrm{d}S = \boldsymbol{F}_{\mathrm{ext}}, \quad (7)$$

$$\int_{\partial P} (\boldsymbol{x} - \boldsymbol{X}) \times (\boldsymbol{\sigma} \cdot \boldsymbol{n}_p) \, \mathrm{d}S = \boldsymbol{0}, \quad (8)$$

where $\boldsymbol{n}_p$ is the outwardly-directed unit normal vector on the boundary of the particle ∂P, $\boldsymbol{X}$ is the position of the center point of the particle, and $\boldsymbol{\sigma}$ is evaluated on the fluid side of the particle boundary ∂P. The translational velocity of the particle is denoted by $\boldsymbol{U}$, which is an unknown in the problem, but will be such that Equation (7) is satisfied. The relation between the particle position and velocity is obtained by the following kinematic equation:

$$\frac{d\boldsymbol{X}}{dt} = \boldsymbol{U}. \quad (9)$$

2.2. *Boundary Conditions*

On the particle boundary and rigid walls S_1 and S_3, a no-slip condition is imposed. On S_1, the velocity is set to zero, whereas on S_3, the velocity is tangential to the curved surface. Periodicity of the surfaces S_2 and S_4 is assumed. Surfaces S_5 and S_6 are assumed to be far enough from the particle to not feel the presence of the particle. Symmetry is prescribed in surfaces S_5 and S_6, i.e., the velocities in the r- and θ-direction are left free, while the velocity in the z-direction is prescribed to be zero.

The velocities on the side curves of surfaces S_5 and S_6 are periodic in the θ- and r-direction. This yields the following boundary conditions:

$$u = U + \omega \times (x - X) \quad \text{on } \partial P, \tag{10}$$

$$u = 0 \quad \text{on } S_1, \tag{11}$$

$$u \cdot e_\theta = U_\mathrm{w} \quad \text{on } S_3, \tag{12}$$

$$u \cdot e_r = u \cdot e_z = 0 \quad \text{on } S_3, \tag{13}$$

$$u \cdot e_z = 0 \quad \text{on } S_5 \text{ and } S_6, \tag{14}$$

$$t \cdot e_r = t \cdot e_\theta = 0 \quad \text{on } S_5 \text{ and } S_6, \tag{15}$$

$$u_4 = R \cdot u_2, \tag{16}$$

$$t_4 = -R \cdot t_2, \tag{17}$$

where ω is the particle angular velocity and the subscripts containing a number denoting the velocities on the surfaces S_1 up to S_6, respectively. Equation (16) imposes a periodic boundary condition for the velocities on surfaces S_2 and S_4. The traction vector on the surface with outwardly-directed normal n is denoted by $t = \sigma \cdot n$. The rotation ϕ of the velocities and tractions from surfaces S_2 to S_4 is performed by the rotation tensor R. More details on the rotation tensor can be found in Appendices A and B. On surface S_6, the solution of the conformation of c_S of the problem without a particle is prescribed since this surface is an inflow surface due to the rigid body movement of the whole mesh with the particle, as will be explained in Section 3.3. This conformation c_S is obtained by solving a separate 1D problem of a Couette flow without the particle and using the steady state value of c_S. This means that for the conformation tensor, the following boundary condition is applied:

$$c = c_\mathrm{S} \quad \text{on } S_6. \tag{18}$$

Furthermore, a periodic boundary condition is also applied for the conformation tensor on surfaces S_2 and S_4. This periodic boundary condition is imposed by the following equation in the same way as is done for the velocities, but now for a tensor instead of a vector:

$$c_2 = R \cdot c_4 \cdot R^T. \tag{19}$$

An initial condition is needed for the conformation tensor c. In order to reduce the computation time, the steady-state solution for the problem without a particle is prescribed as the initial condition for the conformation:

$$c(t = 0) = c_\mathrm{S}|_{t \to \infty}. \tag{20}$$

2.3. *Arbitrary Lagrange–Euler Formulation*

The domain is described with a boundary-fitted mesh that is moved in time in such a way that the mesh moves with the particle, but not necessarily with the fluid. Therefore, the governing equations have to be rewritten in the Arbitrary Lagrange–Euler (ALE) formulation [15]. The movement of the grid has to be compensated. This only applies to the evolution equation of the conformation tensor, since this is the only equation that contains a convective term. The material derivative in Equation (6) is now written as:

$$\frac{Dc}{Dt} = \left.\frac{\partial c}{\partial t}\right|_\xi + (u - u_\mathrm{m}) \cdot \nabla c, \tag{21}$$

where $\partial c/\partial t|_\xi$ denotes the time derivative in a fixed grid point and u_m is the mesh velocity. Details about the mesh velocity will be explained in Section 3.3.

3. Numerical Method

The finite element method is used to solve the problem. For the velocity and the pressure, isoparametric, tetrahedral P_2P_1 (Taylor–Hood) elements are used, whereas for the conformation, tetrahedral P_1 elements are used. In order to solve the constitutive equation, the log-conformation representation [16] and DEVSS-G are used for stability [17].

3.1. Weak Formulation

In this section, the weak formulation of the balance equations is presented. The motion of the particle is enforced as a constraint, using the combined equation of motion approach [18]. The periodic boundary condition for the velocity is also enforced as a constraint, using Lagrange multipliers. The weak form of this problem can now be formulated as follows: find $\boldsymbol{u}$, p, $\boldsymbol{U}$, $\boldsymbol{c}$, $\boldsymbol{\omega}$ and $\boldsymbol{\lambda}$ such that:

$$((\nabla \boldsymbol{v})^T, \nu(\nabla \boldsymbol{u} - \boldsymbol{G}^T)) + (\boldsymbol{D}_v, 2\eta \boldsymbol{D} + \boldsymbol{\tau}_p) - (\nabla \cdot \boldsymbol{v}, p) + \langle \boldsymbol{v} - (\boldsymbol{V} + \boldsymbol{\chi} \times (\boldsymbol{x} - \boldsymbol{X})), \boldsymbol{\lambda} \rangle_{\partial P} = \boldsymbol{V} \cdot \boldsymbol{F}, \quad (22)$$

$$(\boldsymbol{H}, -\nabla \boldsymbol{u} + \boldsymbol{G}^T) = 0, \quad (23)$$

$$(q, \nabla \cdot \boldsymbol{u}) = 0, \quad (24)$$

$$\left(\boldsymbol{d} + \tau(\boldsymbol{u} - \boldsymbol{u}_\mathrm{m}) \cdot \nabla \boldsymbol{d}, \frac{D\boldsymbol{s}}{Dt} - \boldsymbol{g}\Big(\boldsymbol{G}, \boldsymbol{s}\Big)\right) = 0, \quad (25)$$

$$\langle \boldsymbol{\mu}, \boldsymbol{u} - (\boldsymbol{U} + \boldsymbol{\omega} \times (\boldsymbol{x} - \boldsymbol{X})) \rangle_{\partial P} = 0, \quad (26)$$

for all admissible test functions $\boldsymbol{v}, \boldsymbol{H}, q, \boldsymbol{d}, \boldsymbol{\chi}, \boldsymbol{V}, \boldsymbol{\mu}$. Furthermore, $\boldsymbol{D}_v = (\nabla \boldsymbol{v} + (\nabla \boldsymbol{v})^T)/2$, and $(\cdot,\cdot)$ denotes an inner product on domain Ω, whereas $\langle \cdot,\cdot \rangle_{\partial P}$ is an inner product on ∂P; τ and ν are parameters for SUPG and DEVSS-G stabilization, respectively, and $\boldsymbol{s} = \log \boldsymbol{c}$. More information on log-conformation stabilization and the function $\boldsymbol{g}$ can be found in [16], whereas more information on the DEVSS-G method and the projected velocity gradient $\boldsymbol{G}$ can be found in [17]. A finite element method on moving meshes aligned with the particle boundary is used to solve the equations. Since the elements are aligned with the particle boundary, the inner product on ∂P can be calculated in the nodes. This is done using a collocation method [18]:

$$\langle \boldsymbol{\mu}, \boldsymbol{u} - (\boldsymbol{U} + \boldsymbol{\omega} \times (\boldsymbol{x} - \boldsymbol{X})) \rangle_{\partial P} = \sum_{k=1}^{n_\mathrm{coll}} \boldsymbol{\mu}_k \cdot [\boldsymbol{u}(\boldsymbol{x}_k) - (\boldsymbol{U} + \boldsymbol{\omega} \times (\boldsymbol{x}_k - \boldsymbol{X}))], \quad (27)$$

where n_coll is the number of collocation points, which corresponds to the nodes on the particle boundary. Furthermore, $\boldsymbol{\mu}_\mathrm{k}$ and $\boldsymbol{x}_k$ are the Lagrange multiplier and coordinate of the kth collocation point, respectively. The same approach can be used for the weak formulation of the periodic boundary condition on the velocity. Note that the periodic surfaces have to be meshed such that the meshes are also periodic.

3.2. Implementation Periodic Boundary Conditions

In order to impose periodic boundary conditions for the velocity, the velocity has to be prescribed such that the $\boldsymbol{e}_r$, $\boldsymbol{e}_\theta$ and $\boldsymbol{e}_z$ components on S_2 and S_4 are equal. This means that $\boldsymbol{u}_4$ is equal to $\boldsymbol{u}_2$, but rotated. The actual implementation is done in Cartesian coordinates. The same approach is used for the periodic boundary condition for the conformation tensor, only now, a tensor rotation is used instead of a vector rotation. More details on the implementation can be found in Appendices A and B.

3.3. Mesh Movement

Since the mesh will move in time, it is necessary to move the mesh to track the particle boundary. This is done using the ALE method. The starting mesh in the ALE method is used until it becomes too distorted. When this happens, a new mesh, covering the same domain as the old one, is generated using Gmsh [19], and the solution on the old mesh is projected onto the new one. The projection problem is solved to obtain the solution variables on the new mesh. This remeshing and solving the projection problem is done in the same way as was done by Jaensson et al. [20]. However, the projection leads to inaccuracies, and the remeshing procedure is very time consuming. Therefore, the number of remeshing steps should be reduced as much as possible during the simulations.

The mesh velocity $\boldsymbol{u}_{\mathrm{m}}$ can be set to an arbitrary value, and this can be exploited to reduce the remeshing steps, or even avoid remeshing [15]. Therefore, the whole mesh moves as a rigid body with the particle in the θ- and z-direction, while the mesh still deforms in the r-direction due to the movement of the particle. The mesh displacement in the r-direction due to the migration of the particle can be obtained by solving a Laplace equation, in order to guarantee a smooth mesh motion [15].

3.3.1. Prescribing the Conformation Tensor

Since the conformation tensor obeys a hyperbolic partial differential transport equation, the conformation has to be prescribed at inflow boundaries [21,22]. The bottom surface S_6 of the domain can be seen as an inflow surface, since the whole mesh is moved as a rigid body with the particle in the z-direction. Therefore, a solution of the conformation tensor from the same problem without a particle is prescribed at surface S_6.

3.4. *Time Integration*

The time stepping approach now contains the following steps:

Step 1: Compute the particle position at the beginning of every time step. The new particle position can be found using an explicit Euler scheme for Equation (9):

$$\boldsymbol{X}_1 = \boldsymbol{X}_0 + \Delta t \boldsymbol{U}_0. \tag{28}$$

For subsequent time steps, however, a second-order Adams–Bashforth scheme is used:

$$\boldsymbol{X}_{n+1} = \boldsymbol{X}_n + \Delta t(\frac{3}{2}\boldsymbol{U}_n - \frac{1}{2}\boldsymbol{U}_{n-1}). \tag{29}$$

Step 2: Update the mesh, using the new particle position as explained in Section 3.3.

Step 3: The mesh velocities can now be obtained by numerically differentiating the mesh coordinates. For the first time step, the mesh velocity $\boldsymbol{u}_{\mathrm{m}}$ is obtained in each node using a backward differencing scheme:

$$\boldsymbol{u}_{\mathrm{m}} = \frac{\boldsymbol{x}_{\mathrm{m}}^1 - \boldsymbol{x}_{\mathrm{m}}^0}{\Delta t}, \tag{30}$$

where $\boldsymbol{x}$ are the nodal coordinates of the mesh. For subsequent time steps, a second-order backward differencing scheme is used:

$$\boldsymbol{u}_{\mathrm{m}}^{n+1} = \frac{\frac{3}{2}\boldsymbol{x}_{\mathrm{m}}^{n+1} - 2\boldsymbol{x}_{\mathrm{m}}^{n} + \frac{1}{2}\boldsymbol{x}_{\mathrm{m}}^{n-1}}{\Delta t}. \tag{31}$$

The z- and θ-component of the mesh velocity are set equal to the particle velocity in the z- and θ-direction. In this way, the whole mesh moves with the particle.

Step 4: Compute $\boldsymbol{u}^{n+1}$ and p^{n+1}. The method of D'Avino and Hulsen [23] for decoupling the momentum balance from the constitutive equation is applied, using DEVSS-G for stability [17].

Step 5: After solving for the new velocities and pressures, the actual conformation tensor $\boldsymbol{c}^{n+1}$ is found using a second-order, semi-implicit Gear scheme with conformation prediction for Equation (5) using SUPG and the log-conformation representation for stability. The conformation tensor is solved component wise, meaning that not all conformation tensor components are known at the same time. Because of this, it is not possible to enforce periodicity of $\boldsymbol{c}$ on surfaces S_2 and S_4 via a constraint. Therefore, an iteration is performed until the periodic solution for $\boldsymbol{c}$ is found. During this iteration, the conformation tensor has to be described on the inflow-nodes of surfaces S_2 and S_4. See Section 3.3.1. Therefore, it is first computed which of the nodes on surface S_4 are inflow nodes by computing whether $(\boldsymbol{u} - \boldsymbol{u}_\mathrm{m}) \cdot \boldsymbol{n}$ is positive. Here, $\boldsymbol{n}$ is the normal vector of surface S_4. Due to periodicity, outflow nodes on S_2 have to be inflow nodes on S_4.

4. Results

In this section, the convergence of the migration problem will be investigated. After that, the results of the migration of a sedimenting particle without Couette flow, migration of a particle in a Couette flow without sedimentation, and migration of a particle with the two flows combined will be discussed. Finally, the results of the sedimentation of a particle in a Couette flow for two different sets of fluid parameters will be discussed. In order to increase the sedimentation Weissenberg number, the applied downward force on the particle is increased. Since a force is applied on the particle, the sedimentation velocity of the particle is not known a priori. Therefore, the sedimentation Weissenberg number is defined as follows:

$$\Lambda = \lambda \frac{U_\mathrm{N}}{2a}, \tag{32}$$

where U_N is the Stokes velocity defined as $U_\mathrm{N} = F/(6\pi\eta_0 a)$. The shear Weissenberg number is defined as:

$$\mathrm{Wi} = \lambda \frac{U_\mathrm{w}}{R_\mathrm{o} - R_\mathrm{i}}. \tag{33}$$

The ratio between the solvent viscosity and the total viscosity can be expressed as the ratio between the solvent viscosity and the polymer viscosity and can be described as $\beta = \eta_\mathrm{s}/(\eta_\mathrm{p} + \eta_\mathrm{s})$. The viscosity ratio β is chosen to be 0.35, together with a non-linear parameter α of 0.2. The particle is positioned initially at a distance $3a$ from the inner cylinder of the Couette.

The outer radius of the outer cylinder has a value of 90-times the particle radius, while the radius of the inner cylinder has a value of 50-times the particle radius, yielding a gap size of 40-times the particle radius. The height of the Couette flow device is 80-times the particle radius. To save computational costs, only a part of the Couette flow device is modeled, where surface S_4 is rotated over an angle of $\theta = 45^\circ$ with respect to surface S_2. By using $\theta = 45^\circ$, we actually model a Couette flow with eight particles in it, because of the periodic boundary conditions. For this angle, the distance between the particles is larger than 40-times the particle radius, which is large enough to exclude the effect of particle–particle interactions [12].

4.1. Convergence

In this section, the convergence of the migration problem will be investigated. Convergence was the hardest to obtain for a high shear Weissenberg number combined with a high sedimentation Weissenberg number. Therefore, a convergence study was performed for a shear Weissenberg number of $\mathrm{Wi} = 2$ and a sedimentation Weissenberg number of $\Lambda = 2$. In order to choose a mesh with elements small enough that convergence is reached, the migration velocity of a particle, made dimensionless by dividing by a/λ, was plotted for different element sizes h_p around the particle. For these simulations, meshes were constructed with six elements across the gap of the Couette flow device and $h_\mathrm{p} = 20$, $h_\mathrm{p} = 30$, and $h_\mathrm{p} = 40$ elements on the equator of the particle. This element size was extended over a distance a from the particle and then slowly increased to the element size of the surrounding mesh over a distance $9a$. The mesh was refined near the inner radius of the Couette, where 15 elements were placed along the inner radius. The mesh for $h_\mathrm{p} = 30$ is shown in Figure 2. The results for mesh convergence are shown in Figure 3a. Time convergence was studied by plotting again the dimensionless migration velocity of the particle, for a mesh with $h_{\mathrm{p}} = 30$, but now with different time step sizes Δt. The result is shown in Figure 3b. Note that we use quadratic elements for the velocity, meaning that $h_{\mathrm{p}} = 30$ means 61 nodes on the equator of the particle, and six elements across the gap means 13 nodes across the gap.

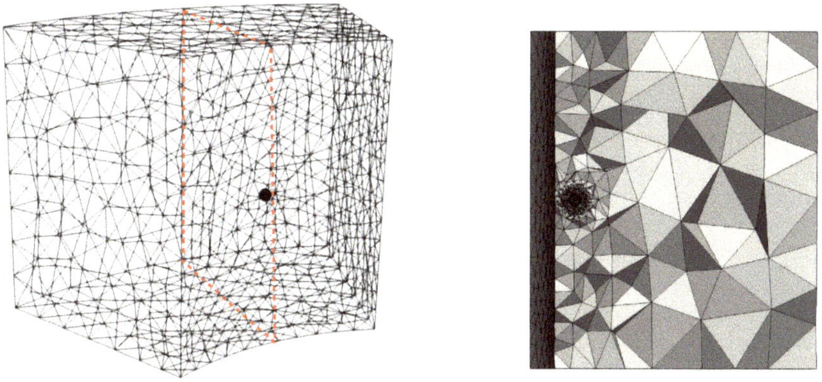

(**a**) Mesh (**b**) Mesh refinement around particle

Figure 2. Mesh that is refined around the particle boundary and at the inner cylinder for $h_{\mathrm{p}} = 30$. Red dotted lines in (**a**) indicate the cross-section S_{CS} shown in (**b**).

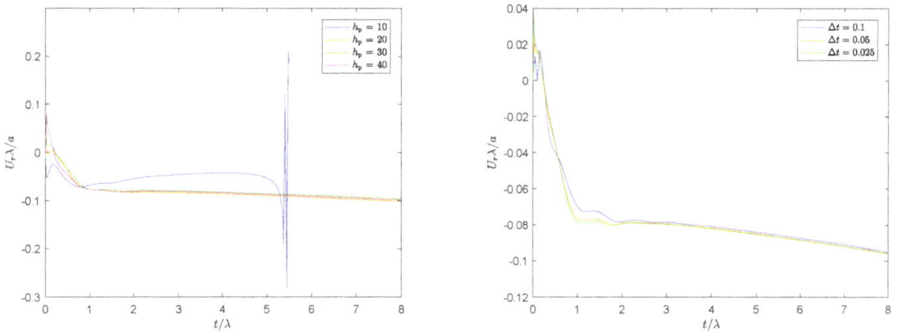

(**a**) Mesh (**b**) Time

Figure 3. Migration velocity of a particle placed at a distance $3a$ from the inner cylinder of a Couette for different element sizes around the particle h_{p} and different time step sizes Δt.

These figures show that for small enough element sizes and time step sizes, the lines superimpose. The convergence is studied here for the most complicated scenario of high Weissenberg numbers and the particle placed originally close to the inner cylinder. Since Figure 3 shows relatively good convergence, we expected that convergence would be better for the less difficult scenarios. In order to have a good balance between accuracy and computational costs, a element size of $h_{\mathrm{p}} = 30$ and a time step size of $\Delta t = 0.05$ were chosen for the simulations.

4.2. Migration

In this section, it is investigated how the two complex flows (shear and sedimentation) influence the migration of a particle placed close to the inner cylinder of a Couette flow device. In order to do this, first the effect of the two separate complex flows is investigated. This is done for different initial particle positions in the Couette flow device, close to the inner cylinder. The dimensionless particle radial position is defined as:

$$\Gamma = \frac{x_{\mathrm{p}} - R_{\mathrm{i}}}{a}, \tag{34}$$

where x_{p} is the radial position of the center of the particle at time $t/\lambda = 10$ and a is the particle radius. Therefore, Γ indicates how many times the particle radius is away from the inner cylinder. The migration velocity at time $t/\lambda = 10$ is used in the figures in this Results Section. For a time that is ten-times the relaxation time of the fluid, it was assumed that the viscoelastic stresses were fully developed, and therefore, all transient effects were due to the migration of the particle. For the representation of the results, the cross-section indicated in red in Figure 2 was used. From here on, this cross-section is referred to as S_{CS}.

4.2.1. Migration for a Falling Particle without Shear in Couette Flow

The migration of a falling particle close to the inner cylinder of the Couette was first investigated without applying a shear flow. This means that the velocity of the outer cylinder $U_{\mathrm{w}} = 0$. The result is shown in Figure 4. Here, the migration velocity U_r of the particle is made dimensionless with a/λ.

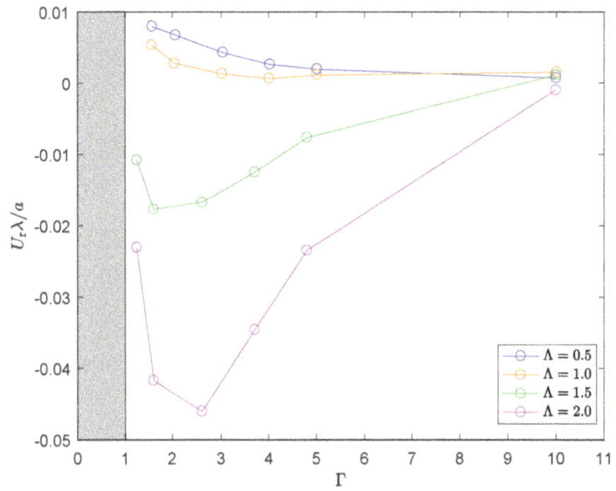

Figure 4. Migration velocity of the particle at different positions from the inner cylinder without shear and varying sedimentation Weissenberg number at time $t/\lambda = 10$. The shaded area indicates the inaccessible region.

This figure shows that the movement of the particle towards the inner cylinder slows down significantly when the particle approaches the inner cylinder. For small values of Λ, it even seems to become positive, i.e., the particle is migrating away from the inner cylinder. This effect of the wall is much more severe for the case of a migrating particle in a Couette flow without sedimentation than for a migrating particle without shear, but with sedimentation, as we will see in Section 4.2.2. The figure also shows that the migration velocity goes to zero when the particle position is further away from the inner cylinder. The movement of the particle towards the inner cylinder speeds up for increasing Λ (i.e., higher negative U_r). The trace of the conformation tensor around the particle for initial particle positions $\Gamma_i = 2$ and $\Gamma_i = 5$ and $\Lambda = 1$ without shear is shown using S_{CS} in Figure 5 at time $t/\lambda = 10$. This figure shows that the trace of the conformation tensor around the particle becomes slightly asymmetric and larger at the side of the particle close to the cylinder, when the particle is positioned close to the inner cylinder.

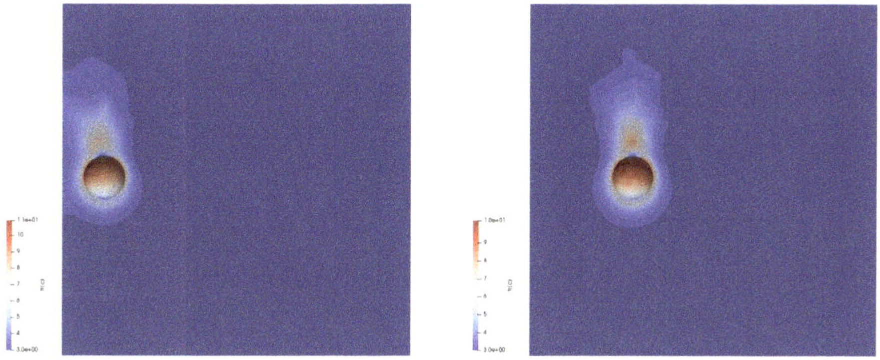

(**a**) $\mathrm{tr}(c)$ for $\Gamma_i = 2$ (**b**) $\mathrm{tr}(c)$ for $\Gamma_i = 5$

Figure 5. The trace of the conformation tensor around the particle for different initial positions from the inner cylinder at time $t/\lambda = 10$ for $\Lambda = 1$ and Wi $= 0$ using S_{CS}.

4.2.2. Migration of a Particle in Couette Flow without Sedimentation

The migration of the particle is now investigated for the case where there is no sedimentation, but the shear rate is varied. This means that the downward force on the particle is set to zero and the velocity of the outer cylinder U_w is varied. The result is shown in Figure 6.

This figure shows that the movement of the particle towards the inner cylinder speeds up with increasing shear Weissenberg number. For low values of Wi, the migration velocity is nearly zero, independent of the position of the particle. For particles close to the inner cylinder, the migration velocity is negative, indicating that the particle migrates towards the inner cylinder. For larger distances from the inner cylinder, the migration velocity is positive for high values of Wi, i.e., the particle seems to migrate away from the inner cylinder. Figure 7 shows the trace of the conformation tensor around the migrating particle with Wi $= 1$ and no sedimentation, for an initial position $\Gamma_i = 2$ and $\Gamma_i = 5$ at time $t/\lambda = 10$ using S_{CS}. This figure shows that the presence of the inner cylinder causes the trace of the conformation to by asymmetric around the particle when the particle is placed close to it. The conformation is higher at the side of the particle close to the inner cylinder, indicating that the polymers are more stretched in the small gap between the particle and the inner cylinder.

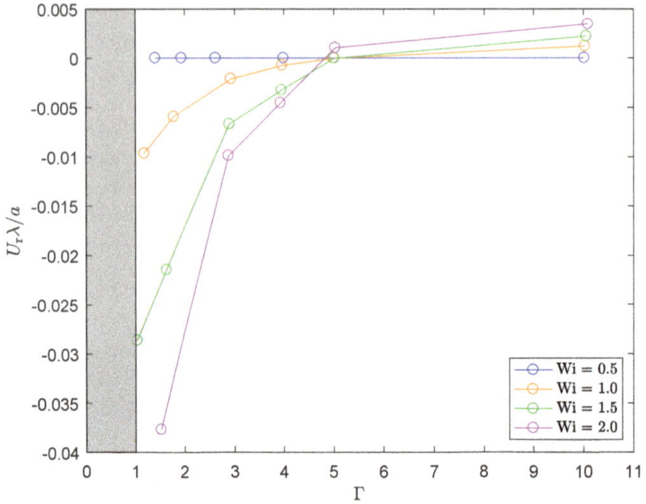

Figure 6. Migration velocity of the particle at different positions from the inner cylinder without sedimentation and varying shear Weissenberg number at time $t/\lambda = 10$. The shaded area indicates the inaccessible region.

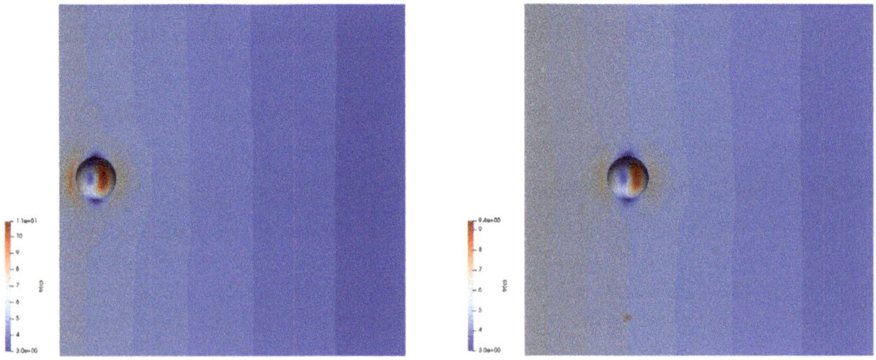

(**a**) $\mathrm{tr}(\boldsymbol{c})$ for $\Gamma_\mathrm{i} = 2$ (**b**) $\mathrm{tr}(\boldsymbol{c})$ for $\Gamma_\mathrm{i} = 5$

Figure 7. The trace of the conformation tensor around the particle for different initial positions from the inner cylinder at time $t/\lambda = 10$ for $\Lambda = 0$ and $\mathrm{Wi} = 1$, using S_{CS}.

4.2.3. Migration of a Falling Particle in a Shear Flow

Now, the two complex flows are combined by first fixing the shear Weissenberg number ($\mathrm{Wi} = 2$) and varying the sedimentation Weissenberg number and vice versa. The results are shown in Figure 8.

Both figures show that when the particle approaches the inner cylinder $\Gamma = 2$, this slows down the movement of the particle towards it for high shear and sedimentation Weissenberg numbers. Here, the particle nearly touches the inner cylinder at $t/\lambda = 10$, drastically reducing the mobility of the particle in the r-direction. Furthermore, it can be observed that combining the two complex flows (solid lines) leads to significantly larger migration velocities than the superposition of the migration velocities of the two separate flows (dashed lines). This effect was more pronounced for large values of both Weissenberg numbers, indicating that for large values of Λ and high Wi, the coupling of the flows becomes non-linear, and therefore, the migration velocity for the combination of the two flows will no longer be a simple superposition of the migration velocities of the two separate flows.

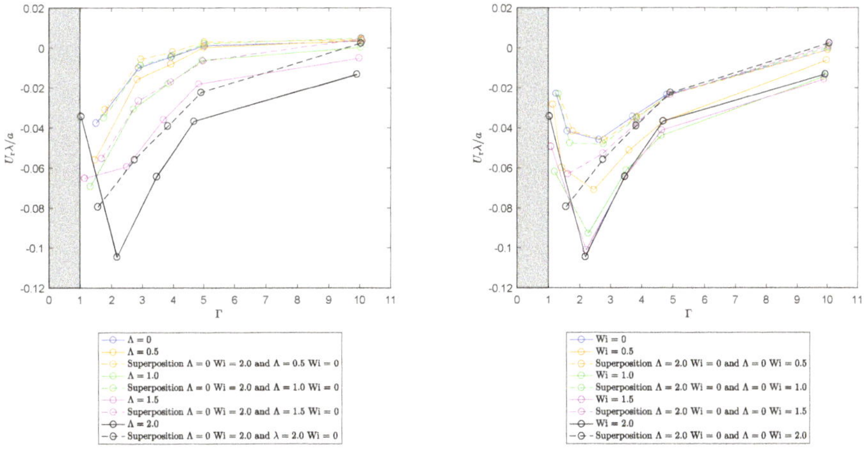

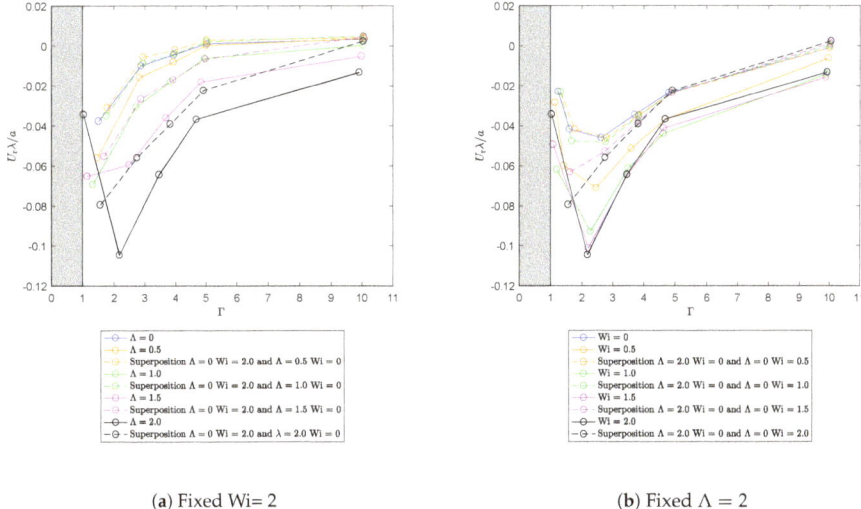

(**a**) Fixed Wi= 2 (**b**) Fixed $\Lambda = 2$

Figure 8. Migration velocity of a particle placed at different initial positions from the inner cylinder with combined complex flows (solid lines) and the superposition of the two separate flows (dashed lines) at time $t/\lambda = 10$. The shaded area indicates the inaccessible region.

Figure 9 shows the trace of the conformation tensor at time $t/\lambda = 10$ for initial particle positions $\Gamma_i = 2$ and $\Gamma_i = 5$ for $\Lambda = 1$ and Wi $= 1$, using S_{CS}. This figure again shows that the trace of the conformation tensor is larger at the side of the particle close to the cylinder, when the particle approaches it. For $\Gamma = 5$ the trace of the conformation tensor is symmetric and high at both sides of the particle due to the shear flow.

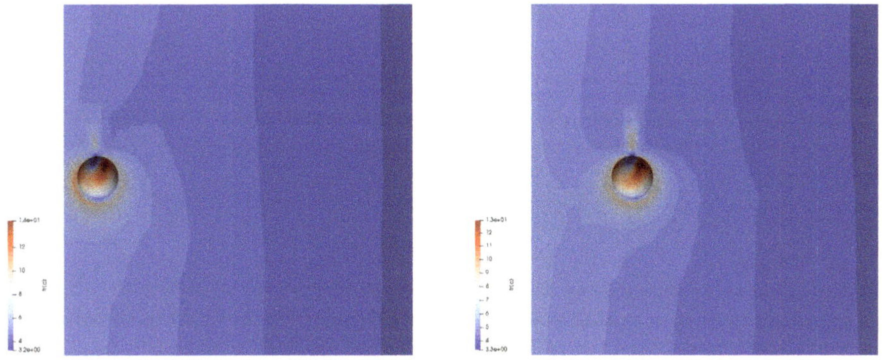

(**a**) $\mathrm{tr}(\boldsymbol{c})$ for $\Gamma_i = 2$ (**b**) $\mathrm{tr}(\boldsymbol{c})$ for $\Gamma_i = 5$

Figure 9. The trace of the conformation tensor around the particle for different initial positions from the inner cylinder at time $t/\lambda = 10$ for $\Lambda = 1$ and Wi $= 1$, using S_{CS}.

Contour plots of the migration velocities for the two flows combined are shown in Figure 10 for $\Gamma = 2$, $\Gamma = 3$, $\Gamma = 4$, $\Gamma = 5$.

These contour plots show, that it is possible to influence the migration velocity of a particle by combining shear and sedimentation flows and changing the shear and sedimentation Weissenberg numbers. Figure 10b for example shows that the migration velocity can be accelerated, when both Weissenberg numbers are high. The migration can be slowed down by lowering one, or both of these Weissenberg numbers.

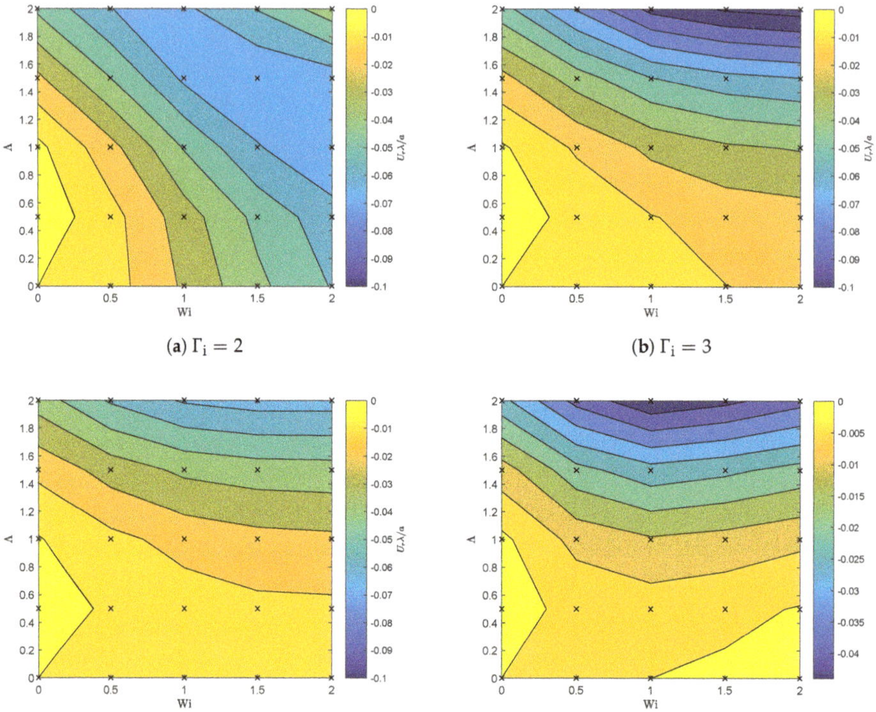

(**a**) $\Gamma_i = 2$ (**b**) $\Gamma_i = 3$

(**c**) $\Gamma_i = 4$ (**d**) $\Gamma_i = 5$

Figure 10. Contour plots of the migration velocities for combined shear and sedimentation flows. Markers x indicate the data points; between these points, the data are interpolated using the MATLAB® function contourf.

4.3. Sedimentation

To investigate the influence of shear and sedimentation Weissenberg number on the sedimentation of a particle, the two complex flows are combined. The particle is now placed initially in the middle of the Couette flow device. Simulations are performed for two different sets of material parameters shown in Table 1. Here, Fluid 1 is used to mimic a Boger fluid with a large value for β and small value for α [24], so the effect of shear thinning can be excluded from the results. Fluid 2 has a much higher value for α, meaning that the fluid is more shear thinning, and a slightly lower value for β. These parameters are the same as for the migration simulations. The zero-shear viscosity $\eta_0 = \eta_s + \eta_p$ is the same for both fluids.

Table 1. Material parameters.

	β	Non-Linear Parameter α
Fluid 1	0.5	0.01
Fluid 2	0.35	0.2

The sedimentation velocity for both fluids, scaled with the sedimentation velocity in a Newtonian fluid with viscosity $\eta_0 = \eta + \eta_p$, is plotted as a function of the shear Weissenberg number for different sedimentation Weissenberg numbers in Figure 11.

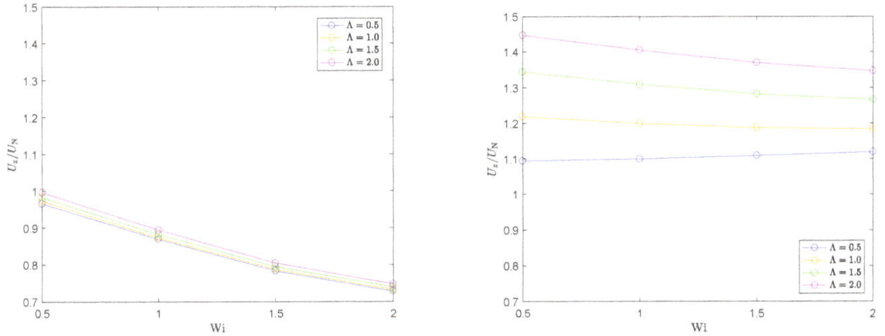

(**a**) Fluid 1 (**b**) Fluid 2

Figure 11. Sedimentation velocity as a function of Wi for increasing Λ for Fluid 1 and Fluid 2.

This figure shows that the sedimentation velocity is indeed decreasing with increasing shear Weissenberg number. This effect is much more pronounced in Fluid 1 than in Fluid 2. Furthermore, the sedimentation velocity is higher in Fluid 2 than in Fluid 1, and for small Wi, the sedimentation velocity is almost equal to the sedimentation velocity of a particle in a Newtonian fluid in Fluid 1, whereas it is higher than the Newtonian velocity in Fluid 2. This confirms that shear thinning is increasing the sedimentation velocity of a falling particle, also decreasing the effect of elasticity that causes the sedimentation velocity to decrease for higher Wi, as was also found in the literature [10,11]. The contour plots of the sedimentation velocity for both fluids are shown in Figure 12.

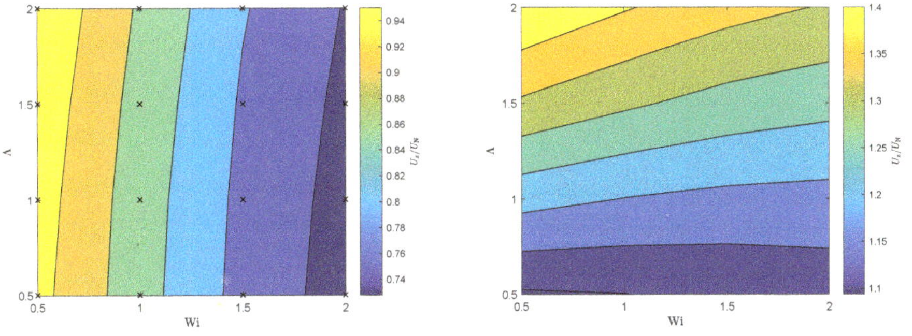

(**a**) Fluid 1 (**b**) Fluid 2

Figure 12. Contour plots of the sedimentation velocity of Fluid 1 and Fluid 2. Markers x indicate the data points; between these points, the data are interpolated using the MATLAB® function contourf.

The contour plots also clearly show different behavior for the two fluids. For Fluid 1, the sedimentation velocity clearly decreases with increasing Wi, while for Fluid 2, the sedimentation velocity seems almost independent of Wi for small Λ, whereas the effect of decreasing velocity for increasing Wi for higher Λ is much less pronounced as for Fluid 1. For Fluid 2, the effect of increasing Λ is also bigger than for Fluid 1.

5. Conclusions

A numerical model was developed in order to investigate numerically the influence of shear and sedimentation Weissenberg number on the migration and sedimentation of a particle in a Couette flow device. DEVSS-G and the log-conformation representation are used for numerical stability. To reduce computational costs, only a part of the Couette flow device is modeled. Periodic boundary conditions are applied at the symmetry planes, using vector and tensor rotations. We believe that this domain excludes the effect of particle–particle interactions. In order to reduce the amount of remeshing steps, the mesh moves as a rigid body with the particle in the θ- and z-direction, while it still deforms due to the movement of the particle in the r-direction. This is done using the ALE method.

The particle was initially placed at different positions close to the inner cylinder of the device. First, the influence of the two separate complex flows was investigated by either applying no velocity U_w on the outer cylinder, and therefore $\mathrm{Wi} = 0$, or applying no downward force on the particle, and therefore $\Lambda = 0$. The suspending fluid is a Giesekus fluid.

First, the migration velocity of a falling particle without applying a shear flow was studied. It was found that the migration velocity was almost zero when the particle was far away from the inner cylinder. When the particle approached the inner cylinder, the particle migrated towards it, and this migration velocity was increased by increasing Λ. At a certain position, the particle got so close to the inner cylinder that the mobility of the particle in the r-direction was drastically decreased, decreasing the migration velocity. Secondly, the migration of a particle in a Couette flow, without sedimentation, was studied. Here, it was found that for small values of Wi, there was no migration. When Wi was increased, the particle started to migrate towards the inner cylinder when it was initially placed close to it. When the particle was initially positioned further away from the cylinder, the particle migrated away from it. When the two flows were combined, it was found that combining the two complex flows leads to a higher migration velocity than the sum of the migration velocities of the two separate flows. This effect was more pronounced for higher Weissenberg numbers, indicating that for high values of Λ and Wi, the coupling of the two flows becomes non-linear. The trace of the conformation tensor was found to be asymmetric around the particle when the particle was in close proximity of the inner cylinder. The trace of the conformation is higher at the side of the particle close to the inner cylinder, indicating that the polymers are more stretched here.

To study the influence of shear and sedimentation Weissenberg number on the sedimentation of a particle, two different sets of material parameters were used, where Fluid 2 is much more shear thinning than Fluid 1 and fluid 1 has a large β, mimicking a Boger fluid. It was confirmed that shear thinning increases the sedimentation velocity of the particle. Furthermore, it was found that the sedimentation velocity indeed decreases for increasing Wi in Fluid 1, whereas this effect was less pronounced for Fluid 2. For small values of Λ, this effect was not seen at all for Fluid 2, indicating that shear thinning opposes the effect of elasticity.

Author Contributions: This study was a part of the graduation project of M.M.A.S. She was responsible for developing the computational model and for performing the simulations and data analysis. She was supervised by N.O.J., who also contributed to the interpretation of the results. N.O.J., M.A.H. and P.D.A. contributed to the critical revision of the article.

Funding: This research received no external funding.

Conflicts of Interest: The authors declare no conflict of interest.

Appendix A. Periodic Boundary Condition Velocity

A schematic overview of the domain in 2D is shown in Figure A1. The domain is shown in 2D for clarity, because the velocity in the z-direction does not undergo a rotation. The z-component is now the vector pointing into the paper.

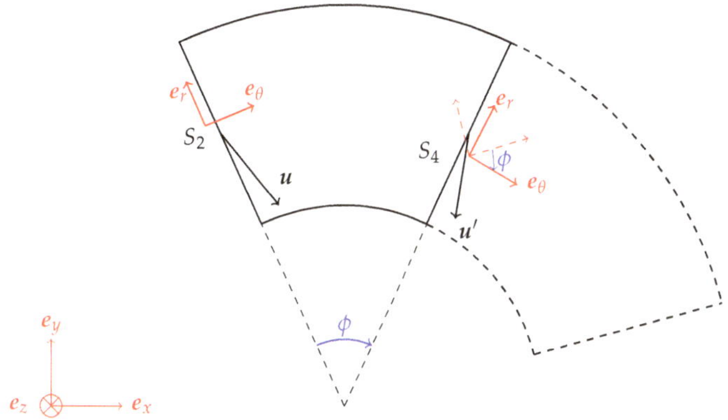

Figure A1. Schematic overview of a 2D curve with periodic boundary conditions.

To obtain the velocity components on surface S_2 equal to the velocity components on S_4, $\boldsymbol{u}_2$ has to be rotated over a counterclockwise angle ϕ, since surface S_4 is rotated over an angle ϕ with respect to surface S_2. The rotation of a vector can be performed by taking the dot product with a tensor $\boldsymbol{R}$, which rotates the vector in a constant coordinate system. After applying the rotation, the velocity on surface S_2 equals the velocity on S_4. In a Cartesian coordinate system ($\boldsymbol{e}_x$, $\boldsymbol{e}_y$, $\boldsymbol{e}_z$), the rotation is expressed as follows:

$$\begin{pmatrix} U_x \\ U_y \\ U_z \end{pmatrix} = \begin{pmatrix} \cos\phi & -\sin\phi & 0 \\ \sin\phi & \cos\phi & 0 \\ 0 & 0 & 1 \end{pmatrix} \begin{pmatrix} U'_x \\ U'_y \\ U'_z \end{pmatrix}. \tag{A1}$$

Using Equation (A1), U_x, U_y, and U_z on surface S_4 can be expressed as follows:

$$U_x = U'_x \cos\phi - U'_y \sin\phi, \tag{A2}$$

$$U_y = U'_x \sin\phi + U'_y \cos\phi, \tag{A3}$$

$$U_z = U'_z. \tag{A4}$$

The same approach can be used to calculate the traction components on surfaces S_4 and S_2, in order to obey the periodic boundary conditions.

Appendix B. Periodic Boundary Condition Conformation Tensor

Appendix B.1. Tensor Rotation

Periodic boundary conditions for the conformation tensor are needed on surfaces S_2 and S_4, since only a part of the Couette flow device is modeled. This is done following the same principle as for the periodic boundary conditions for the velocity, only now a tensor rotation has to be performed instead of a vector rotation. The formula to rotate a tensor $\boldsymbol{A}$ can be derived as follows:

The rotation of an arbitrary vector $\boldsymbol{u}'$ can be performed by taking the product of a tensor that performs a rotation with this vector. Now, the rotated vector $\boldsymbol{u}$ of $\boldsymbol{u}'$ is obtained:

$$\boldsymbol{u} = \boldsymbol{R} \cdot \boldsymbol{u}', \tag{A5}$$

where $\boldsymbol{R}$ is a tensor that performs a rotation in a constant coordinate system. Furthermore, the inner product of a rotated tensor $\boldsymbol{A}$ with a rotated vector $\boldsymbol{x}$ gives a rotated vector $\boldsymbol{b}$, and the same thing holds for the unrotated tensor and vectors:

$$\boldsymbol{A} \cdot \boldsymbol{x} = \boldsymbol{b} \tag{A6}$$

$$\boldsymbol{A}' \cdot \boldsymbol{x}' = \boldsymbol{b}'. \tag{A7}$$

If these equations are combined with Equation (A5), the following can be obtained:

$$\boldsymbol{A} \cdot \boldsymbol{R} \cdot \boldsymbol{x}' = \boldsymbol{R} \cdot \boldsymbol{b}'. \tag{A8}$$

Equation (A8) can be multiplied with $\boldsymbol{R}^{-1}$, $(\boldsymbol{R}^{-1} \cdot \boldsymbol{R} = \boldsymbol{I})$, to obtain:

$$\boldsymbol{R}^{-1} \cdot \boldsymbol{A} \cdot \boldsymbol{R} \cdot \boldsymbol{x}' = \boldsymbol{b}'. \tag{A9}$$

Equation (A9) combined with Equation (A7) states that $\boldsymbol{A}' = \boldsymbol{R}^{-1} \cdot \boldsymbol{A} \cdot \boldsymbol{R}$. If $\boldsymbol{A}' = \boldsymbol{R}^{-1} \cdot \boldsymbol{A} \cdot \boldsymbol{R}$ is pre-multiplied with $\boldsymbol{R}$ and post-multiplied with $\boldsymbol{R}^{-1}$, we obtain: $\boldsymbol{R} \cdot \boldsymbol{A}' \cdot \boldsymbol{R}^{-1} = \boldsymbol{R} \cdot \boldsymbol{R}^{-1} \cdot \boldsymbol{A} \cdot \boldsymbol{R} \cdot \boldsymbol{R}^{-1}$. Invoking that $\boldsymbol{R}$ is an orthogonal tensor ($\boldsymbol{R}^T = \boldsymbol{R}^{-1}$), the rotation of tensor $\boldsymbol{A}'$ to tensor $\boldsymbol{A}$ can be performed as follows:

$$\boldsymbol{A} = \boldsymbol{R} \cdot \boldsymbol{A}' \cdot \boldsymbol{R}^{-1} = \boldsymbol{R} \cdot \boldsymbol{A}' \cdot \boldsymbol{R}^T. \tag{A10}$$

Appendix B.2. Conformation Tensor

Periodic boundary conditions for the conformation tensor are applied to surfaces S_2 and S_4. This means that the components c_{xx}, c_{xy}, c_{xz}, c_{yx}, c_{yy}, c_{yz}, c_{zx}, c_{zy}, and c_{zz} on the inflow nodes of surfaces S_2 and S_4 have to be equal to the components c'_{xx}, c'_{xy}, c'_{xz}, c'_{yx}, c'_{yy}, c'_{yz}, c'_{zx}, c'_{zy}, and c'_{zz} on the outflow nodes of surfaces S_2 and S_4, but rotated. Since the surface S_2 is rotated over an angle ϕ compared to surface S_4, the components of $\boldsymbol{c}'$ on the inflow nodes have to be rotated over the same angle ϕ. Using the result of Equation (A10), this gives the following equation for rotating the conformation tensor $\boldsymbol{c}'$:

$$\begin{pmatrix} c_{xx} & c_{xy} & c_{xz} \\ c_{yx} & c_{yy} & c_{yz} \\ c_{zx} & c_{zy} & c_{zz} \end{pmatrix} = \begin{pmatrix} \cos\phi & -\sin\phi & 0 \\ \sin\phi & \cos\phi & 0 \\ 0 & 0 & 1 \end{pmatrix} \begin{pmatrix} c'_{xx} & c'_{xy} & c'_{xz} \\ c'_{yx} & c'_{yy} & c'_{yz} \\ c'_{zx} & c'_{zy} & c'_{zz} \end{pmatrix} \begin{pmatrix} \cos\phi & \sin\phi & 0 \\ -\sin\phi & \cos\phi & 0 \\ 0 & 0 & 1 \end{pmatrix}. \tag{A11}$$

The equations for the tensor components can be expressed using Equation (A11):

$$c_{xx} = (c'_{xx}\cos\phi - c'_{yx}\sin\phi)\cos\phi - \sin\phi(c'_{xy}\cos\phi - c'_{yy}\sin\phi) \tag{A12}$$

$$c_{xy} = (c'_{xx}\cos\phi + c'_{yx}\sin\phi)\sin\phi + \cos\phi(c'_{xy}\cos\phi - c'_{yy}\sin\phi) \tag{A13}$$

$$c_{xz} = c'_{xz}\cos\phi - c'_{yz}\sin\phi \tag{A14}$$

$$c_{yx} = (c'_{xx}\sin\phi + c'_{yx}\cos\phi)\cos\phi - \sin\phi(c'_{xy}\sin\phi + c'_{yy}\cos\phi) \tag{A15}$$

$$c_{yy} = (c'_{xx}\sin\phi + c'_{yx}\cos\phi)\sin\phi + \cos\phi(c'_{xy}\sin\phi + c'_{yy}\cos\phi) \tag{A16}$$

$$c_{yz} = c'_{xz}\sin\phi + c'_{yz}\cos\phi \tag{A17}$$

$$c_{zx} = c'_{zx} \cos\phi - c'_{zy} \sin\phi \tag{A18}$$

$$c_{zy} = c'_{zx} \sin\phi + c'_{zy} \cos\phi \tag{A19}$$

$$c_{zz} = c'_{zz} \tag{A20}$$

Since, $c_{xy} = c_{yx}$, $c_{xz} = c_{zx}$, and $c_{yz} = c_{zy}$, it can be concluded that the conformation tensor is symmetric.

References

1. Kim, Y.; Yoo, J. Three-dimensional focusing of red blood cells in microchannel flows for bio-sensing applications. *Biosens. Bioelectron.* **2009**, *24*, 3677–3682. [CrossRef] [PubMed]
2. Economides, M.; Nolte, K. *Reservoir Stimulation*; Prentice Hall: Hoboken, NJ, USA, 1989.
3. Singh, P.; Joseph, D. Sedimentation of a sphere near a vertical wall in an Oldroyd-B fluid. *J. Non-Newton. Fluid Mech.* **2000**, *94*, 179–203. [CrossRef]
4. D'Avino, G.; Tuccillo, T.; Maffettone, P.; Greco, F.; Hulsen, M. Numerical simulations of particle migration in a viscoelastic fluid subjected to shear flow. *Comput. Fluids* **2010**, *39*, 709–721. [CrossRef]
5. D'Avino, G.; Snijkers, F.; Pasquino, R.; Hulsen, M.; Greco, F.; Maffettone, P.; Vermant, J. Migration of a sphere suspended in viscoelastic liquids in Couette flow: Experiments and simulations. *Rheol. Acta* **2012**, *51*, 215–234. [CrossRef]
6. Chilcott, M.; Rallison, J. Creeping flow of dilute polymer solutions past cylinders and spheres. *J. Non-Newton. Fluid Mech.* **1988**, *29*, 381–432. [CrossRef]
7. Harlen, O.; Rallison, J.; Chilcott, M. High-Deborah-number flows of dilute polymer solutions. *J. Non-Newton. Fluid Mech.* **1990**, *34*, 319–349. [CrossRef]
8. Arigo, M.; Rajagopalan, D.; Shapley, N.; McKinley, G. The sedimentation of a sphere through an elastic fluid. Part 1. Steady motion. *J. Non-Newton. Fluid Mech.* **1995**, *60*, 225–27. [CrossRef]
9. D'Avino, G.; Maffettone, P. Particle dynamics in viscoelastic liquids. *J. Non-Newton. Fluid Mech.* **2015**, *215*, 80–104. [CrossRef]
10. Van den Brule, B.; Gheissary, G. Effects of fluid elasticity on the static and dynamic settling of a spherical particle. *J. Non-Newton. Fluid Mech.* **1993**, *49*, 123–132. [CrossRef]
11. Padhy, S.; Rodriguez, M.; Shaqfeh, E.; Iaccarino, G.; Morris, J.; Tonmakayakul, N. The effect of shear thinning and walls on the sedimentation of a sphere in an elastic fluid under orthogonal shear. *J. Non-Newton. Fluid Mech.* **2013**, *201*, 120–129. [CrossRef]
12. Padhy, S.; Shaqfeh, E.; Iaccarino, G.; Morris, J.; Tonmakayakul, N. Simulations of a sphere sedimenting in a viscoelastic fluid with cross shear flow. *J. Non-Newton. Fluid Mech.* **2013**, *197*, 48–60. [CrossRef]
13. Murch, W.; Krishnan, S.; Shaqfeh, E.; Iaccarino, G. Growth of viscoelastic wings and the reduction of particle mobility in a viscoelastic shear flow. *Phys. Rev. Fluids* **2017**, *2*, 103302. [CrossRef]
14. Giesekus, H. A simple constitutive equation for polymer fluids based on the concept of deformation-dependent tensorial mobility. *J. Non-Newton. Fluid Mech.* **1982**, *11*, 69–109. [CrossRef]
15. Hu, H.; Patankar, N.; Zhu, M. Direct numerical simulations of fluid-solid systems using the arbitrary Lagrangian-Eulerian technique. *J. Comput. Phys.* **2001**, *169*, 427–462. [CrossRef]
16. Hulsen, M.; Fattal, R.; Kupferman, R. Flow of viscoelastic fluids past a cylinder at high Weissenberg number: Stabilized simulations using matrix logarithms. *J. Non-Newton. Fluid Mech.* **2005**, *127*, 27–39. [CrossRef]
17. Bogaerds, A.; Grillet, A.; Peters, G.; Baaijens, F. Stability analysis of polymer shear flows using the extended pom-pom constitutive equations. *J. Non-Newton. Fluid Mech.* **2002**, *108*, 187–208. [CrossRef]
18. Glowinski, R.; Pan, T.; Hesla, T.; Joseph, D. A distributed Lagrange multiplier/fictitious domain method for particulate flows. *Int. J. Multiph. Flow* **1999**, *25*, 755–794. [CrossRef]
19. Geuzaine, C.; Remacle, J. Gmsh: A 3D finite element mesh generator with built-in pre- and post-processing facilities. *Int. J. Numer. Methods Eng.* **2009**, *79*, 1–24. [CrossRef]
20. Jaensson, N.; Hulsen, M.; Anderson, P. Stokes-Cahn-Hilliard formulations and simulations of two-phase flows with suspended rigid particles. *Comput. Fluids* **2015**, *111*, 1–17. [CrossRef]
21. Van der Zanden, J.; Hulsen, M. Mathematical and physical requirements for successful computations with viscoelastic fluid models. *J. Non-Newton. Fluid Mech.* **1988**, *29*, 93–117. [CrossRef]

22. Renardy, M. Inflow boundary conditions for steady flows of viscoelastic fluids with differential constitutive laws. *J. Non-Newton. Fluid Mech.* **1988**, *18*, 445–454.
23. D'Avino, G.; Hulsen, M. Decoupled second-order transient schemes for the flow of viscoelastic fluids without a viscous solvent contribution. *J. Non-Newton. Fluid Mech.* **2010**, *165*, 1602–1612. [CrossRef]
24. Snijkers, F.; D'Avino, G.; Maffettone, P.; Greco, F.; Hulsen, M.; Vermant, J. Effect of viscoelasticity on the rotation of a sphere in shear flow. *J. Non-Newton. Fluid Mech.* **2011**, *166*, 363–372. [CrossRef]

fluids

MDPI

Article

DEM/CFD Simulations of a Pseudo-2D Fluidized Bed: Comparison with Experiments

Ziad Hamidouche [1], Yann Dufresne [2], Jean-Lou Pierson [1,*], Rim Brahem [1], Ghislain Lartigue [2] and Vincent Moureau [2]

[1] IFP Energie Nouvelles, Rond-Point de l'échangeur de Solaize, BP 3, 69360 Solaize, France; hamidouche_ziad@yahoo.com (Z.H.); rim.brahem@ifpen.fr (R.B.)
[2] CORIA, CNRS, INSA & Université de Rouen, 76801 Saint-Etienne-du-Rouvray, France; yann.dufresne@insa-rouen.fr (Y.D.); ghislain.lartigue@coria.fr (G.L.); vincent.moureau@coria.fr (V.M.)
* Correspondence: jean-lou.pierson@ifpen.fr

Received: 12 February 2019; Accepted: 4 March 2019; Published: 15 March 2019

Abstract: The present work investigates the performance of a mesoscopic Lagrangian approach for the prediction of gas-particle flows under the influence of different physical and numerical parameters. To this end, Geldart D particles with 1 mm diameter and density of 2500 kg/m^3 are simulated in a pseudo-2D fluidized bed using a Discrete Element Method (DEM)/Large-Eddy Simulation (LES) solver called YALES2. Time-averaged quantities are computed and compared with experimental results reported in the literature. A mesh sensitivity analysis showed that better predictions regarding the particulate phase are achieved when the mesh is finer. This is due to a better description of the local and instantaneous gas-particle interactions, leading to an accurate prediction of the particle dynamics. Slip and no slip wall conditions regarding the gas phase were tested and their effect was found negligible for the simulated regimes. Additional simulations showed that increasing either the particle-particle or the particle-wall friction coefficients tends to reduce bed expansion and to initiate bubble formation. A set of friction coefficients was retained for which the predictions were in good agreement with the experiments. Simulations for other Reynolds number and bed weight conditions are then carried out and satisfactory results were obtained.

Keywords: DEM/CFD simulations; Euler/Lagrange approach; fluidized beds; frictional effects

1. Introduction

Based on their effectiveness regarding gas-solid heat and mass transfers, fluidized beds are among the best options for developing economically and environmentally viable techniques for fossil-fuel-based energy generation. Such systems involve complex physical mechanisms such as momentum, heat and mass exchanges between the gas and the particulate phases. In addition, fluidized beds exhibit an unsteady and inhomogeneous behavior leading to wide characteristic length and time scales. Particle-particle and gas-particle interactions at the micro-scale (1 to $5d_p$, where d_p is the characteristic length of a particle) result in meso-scale structures, such as bubbles (10 to $100d_p$), which can affect the macro-scale gas-particle flow [1]. Further, different local behaviors can be observed depending on the local particle-particle, gas-particle and particle-wall interactions, and may profoundly modify the bed hydrodynamics. Therefore, various numerical approaches have been developed over the past decades to simulate those flows at microscopic, mesoscopic and macroscopic scales with the aim of elucidating the mechanisms underlying the origin and the evolution of the heterogeneous gas-particle flow pattern. The accurate prediction of the underlying physics makes possible to improve existing processes and to design more efficient new facilities. In this context, the development of reliable numerical approaches is an essential prerequisite.

Discrete Element Method (DEM) is among the most appropriate meso-scale approaches to simulate small scale fluidized beds, with $\mathcal{O}(10^6)$ particles [2]. In this technique the particle motion is given by the Newtonian equations. Particles interact with each others through collisions, that can be described using models from molecular dynamics [3], the so-called sphere models [4–6]. In this work the particles are considered as soft spheres that can overlap slightly and exert both normal and tangential forces on each other [4]. This model, known as the soft-sphere model, requires a contact force model in order to account for the inter-particle collision dynamics. The gas flow is solved on an Eulerian grid using a continuum approach based on a volume averaging of the Navier-Stokes equations [7]. Interphase momentum transfer terms are included in the modeling to account for the fluid-particle interactions [8]. In this work, the ability of a DEM/LES approach to reproduce the gas-particle flow behavior in a 2D-fluidized bed is assessed. A specific attention will be paid to the influence of numerical (grid cell size) and physical parameters (friction coefficient).

In Section 2, the modeling strategy is given. Detailed models, based on averaged Navier-Stokes equations for the gas phase, DEM technique for the particles and coupling procedure between the phases are provided in Sections 2.1–2.3, respectively. Detailed information about the numerical schemes can be found in Section 2.4. Finally, an overview of the simulation cases is presented in Section 3 and results discussed in Section 4.

2. Modeling Strategy

All the numerical simulations presented in this work are performed using the finite-volume code YALES2 [9], a Large-Eddy Simulation (LES) and Direct Numerical Simulation (DNS) solver based on unstructured meshes. This code solves the low-Mach number Navier-Stokes equations for turbulent reactive flows using a time-staggered projection method for constant [10] or variable density flows [11]. YALES2 is specifically tailored for solving these low-Mach number equations on massively parallel machines with billion-cell meshes thanks to a highly optimized linear solvers [12].

Recently, a meso-scale four-way coupling approach for the modeling of solid particles has been implemented in the YALES2 solver. This approach combines DEM approach to represent the solid phase with LES equations solved on an Eulerian unstructured grid for the fluid phase. The CFD/DEM solver has been thoroughly optimized for massively parallel computing. It features a dynamic collision detection grid for unstructured meshes and packing/unpacking of the halo data for non-blocking Message Passing Interface (MPI) exchanges.

2.1. Gas Phase Modeling

The governing equations for the averaged fluids flow are obtained from the filtering of the unsteady, low-Mach number Navier-Stokes equations, taking the local fluid and solid fractions into account. If G is a filtering kernel (see for instance [7]), the local fluid fraction ε is defined as:

$$\varepsilon(\mathbf{x},t) = \int_{V_f} G(|\mathbf{x}-\mathbf{y}|)d\mathbf{y}, \tag{1}$$

where V_f is the volume occupied by the fluid. Defining $\Phi(\mathbf{x},t)$ as a function of position and time, the volume filtered field $<\Phi>(\mathbf{x},t)$ refers to the regular spatial average and is computed by taking the convolution product with the filtering kernel G, giving:

$$\varepsilon <\Phi> (\mathbf{x},t) = \int_{V_f} \Phi(\mathbf{y},t)G(|\mathbf{x}-\mathbf{y}|)d\mathbf{y}. \tag{2}$$

Further details concerning the volume filtering operations can be found in [8]. The governing equations finally read:

$$\frac{\partial \varepsilon}{\partial t} + \nabla \cdot (\varepsilon < \mathbf{u} >) = 0, \tag{3}$$

$$\rho\frac{\partial}{\partial t}(\varepsilon < \mathbf{u} >) + \rho\nabla\cdot(\varepsilon < \mathbf{u} > \otimes < \mathbf{u} >) = -\nabla < P > + \nabla\cdot(\varepsilon < \boldsymbol{\tau} >) + \varepsilon\rho\mathbf{g} + \frac{1}{\Delta V} < \mathbf{F} >_{p\rightarrow f}, \quad (4)$$

$\mathbf{u}$, ρ, P, and $\boldsymbol{\tau}$ are the gas velocity, density, dynamic pressure respectively and viscous stress tensor. The closure used to calculate the turbulent viscosity μ_t is the localized dynamic Smagorinsky model [13] proposed by [14,15]. The term $\bar{\mathbf{F}}_{p\rightarrow f}$ is the momentum source term due to particle displacement. ΔV is the local control volume. Details concerning the computation of this term can be found in Section 2.3. For the sake of clarity, the notation $<>$ will be dropped for the averaged quantities in the following.

The use of the LES model for CFD/DEM simulations in case of dense granular flows is questionable. Indeed, there are very few publications dealing with the effect of inertial particles on a turbulent flow in a confined domain. It is thus very difficult to conclude whether the actual flow is turbulent or not. Nevertheless, previous publications (see, for instance, [8]) used this kind of model in fluidized bed simulations. In our case the bulk Reynolds number based on the bed width is relatively high ($5600 \leq Re \leq 8000$) so that, a single gas-phase flow in the same condition, is likely to be turbulent or at least in a transition regime. We performed a series of additional computations in order to assess the effect of the turbulent viscosity on the bed hydrodynamics. Results showed that, both qualitatively and quantitatively, such a model did not change significantly the hydrodynamics of the fluidized bed (see Appendix A). This tends to confirm that in the present configuration, the global behaviour of the flow is driven by the largest scales and by particle-particle contact.

2.2. Discrete Particle Modeling

The forces acting on a particle in motion can be divided into two categories, volume and surface forces. The unique volume force in the present situation is the weight whereas, the surface forces consist of hydrodynamic and contact forces. While the hydrodynamic forces arise from fluid-particle interactions, such as drag and lift forces, the contact forces are due to particle-particle and particle-wall interactions. Such forces can be further classified in collision and adhesive forces. In the present work, due to the high solid/gas density ratio, as reported in Table 1, buoyancy, lift and added mass forces are neglected. To date, theoretical expressions for the history force in dense regimes are still missing. Therefore, such a force is not taken into account in our modeling. Concerning the contact forces, only collision forces are accounted for. Adhesive effects that may originate from electrostatic or Van Der Vaals forces can safely be here neglected since, as it can be seen in Table 1, the simulated particles are relatively large (Geldart D particles).

Discrete particle models, in which each particle is tracked is tracked in a Lagrangian fashion, have clearly shown their ability to simulate the behavior of granular flows, and originate from molecular dynamics methods initiated in the 1950s [3]. A soft-sphere model [16] is employed to compute contact between each particle. They are allowed to overlap other particles or walls in a controlled manner. A resulting contact force accounting for particle-particle and particle-wall repulsion is thus added in the momentum balance of each particle. Particle movement is then given by Newton's second law for translation assuming high solid/gas density ratio:

$$m_p\frac{d\mathbf{u}_p}{dt} = \mathbf{F}_D + \mathbf{F}_G + \mathbf{F}_P + \mathbf{F}_C \text{ with } \frac{d\mathbf{x}_p}{dt} = \mathbf{u}_p, \quad (5)$$

where m_p, $\mathbf{u}_p$, and $\mathbf{x}_p$ are the particle mass, velocity, position respectively. $\mathbf{F}_D$ is the drag force, $\mathbf{F}_G$ is the gravity force, $\mathbf{F}_P$ is the pressure gradient force and $\mathbf{F}_C$ is the collision force. The relation between $\mathbf{F}_D$, $\mathbf{F}_C$ in Equation (5) and $\mathbf{F}_{p\rightarrow f}$ in Equation (4) is detailed in Section 2.3. Preliminary simulations including particle rotation provided similar results as those without rotation. As a consequence, for computational reason, the particle rotation is not accounted for in this paper.

2.2.1. Modeling of Collisions

The total collision force $\mathbf{F}_C$ acting on particle a is computed as the sum of all forces $\mathbf{f}^{col}_{b\to a}$ exerted by the N_p particles and N_w walls in contact. As particles and walls are treated similarly during collisions, the b index refers to both:

$$\mathbf{F}_C = \sum_{b=1}^{N_p+N_w} \mathbf{f}^{col}_{b\to a}, \tag{6}$$

$$\mathbf{f}^{col}_{b\to a} = \mathbf{f}^{col}_{n,b\to a} + \mathbf{f}^{col}_{t,b\to a}. \tag{7}$$

Depending on the desired compromise between accuracy and numerical cost, a lot of soft sphere models can be found in the literature. Here, a linear-spring/dashpot [16] is used along with a simple Coulomb sliding model accounting for the normal $\left(\mathbf{f}^{col}_{n,b\to a}\right)$ and tangential $\left(\mathbf{f}^{col}_{t,b\to a}\right)$ components of the contact force, respectively, as in the work of Capecelatro [8]. For one particle (or wall) b acting on a particle a:

$$\mathbf{f}^{col}_{n,b\to a} = \begin{cases} -k_n\delta_{ab}\mathbf{n}_{ab} - 2\gamma_n M_{ab}\mathbf{u}_{ab,n} & \text{if } \delta_{ab} > 0, \\ 0 & \text{else,} \end{cases} \tag{8}$$

$$\mathbf{f}^{col}_{t,b\to a} = -\mu_{tan}||\mathbf{f}^{col}_{n,b\to a}||\mathbf{t}_{ab}. \tag{9}$$

The term δ_{ab} is defined as the overlap between the a and b entities expressed using each particle radius r_p and center coordinates $\mathbf{x}_p$ such as:

$$\delta_{ab} = r_a + r_b - (\mathbf{x}_b - \mathbf{x}_a).\mathbf{n}_{ab}. \tag{10}$$

The system effective mass M_{ab} is expressed using each particle mass m_p such as:

$$M_{ab} = \frac{1}{1/m_a + 1/m_b}. \tag{11}$$

In case of a particle-wall collision, the wall is considered as a particle with infinite mass and null radius. This model requires three user-defined parameters; k_n, γ_n, and μ_{tan} respectively accounting for the spring stiffness, normal damping, and friction coefficient of the $a-b$ binary system. $\mathbf{n}_{ab}$ and $\mathbf{t}_{ab}$ respectively account for the unit normal vector from particle a towards entity b and the unit tangential vector. $\mathbf{n}_{ab}$ is calculated as follows:

$$\mathbf{n}_{ab} = \frac{\mathbf{x}_b - \mathbf{x}_a}{||\mathbf{x}_b - \mathbf{x}_a||}. \tag{12}$$

with $\mathbf{u}_{ab}$ being the relative velocity of the colliding system, its normal and tangential components are then given by:

$$\mathbf{u}_{ab,n} = ((\mathbf{u}_a - \mathbf{u}_b).\mathbf{n}_{ab})\mathbf{n}_{ab}, \tag{13}$$

$$\mathbf{u}_{ab,t} = \mathbf{u}_{ab} - \mathbf{u}_{ab,n}. \tag{14}$$

Finally, $\mathbf{t}_{ab}$ is given by:

$$\mathbf{t}_{ab} = \frac{\mathbf{u}_{ab,t}}{||\mathbf{u}_{ab,t}||}. \tag{15}$$

Using Newton's third law and a projection on $\mathbf{n}_{ab}$ yields the following ordinary differential equation for the overlap evolution of an $a-b$ binary system undergoing collision without taking any other force into account:

$$\frac{d^2\delta_{ab}}{dt^2} + 2\gamma_n\frac{d\delta_{ab}}{dt} + \omega_0^2\delta_{ab} = 0, \tag{16}$$

where ω_0 stands for the system's natural frequency and is defined as:

$$\omega_0^2 = \frac{k_n}{M_{ab}}. \tag{17}$$

The damping parameter γ_n accounting for the energy dissipation occurring during contact is calculated by the mean of a normal restitution coefficient e_n verifying $0 < e_n \leq 1$ such as:

$$\gamma_n = -\frac{\omega_0 \ln e_n}{\sqrt{\pi^2 + (\ln e_n)^2}}. \tag{18}$$

A contact time T_C can also be analytically extracted from Equation (16) corresponding to the time during which the particle a and the entity b are overlapping:

$$T_C = \frac{\pi}{\sqrt{\omega_0^2 - \gamma_n^2}}. \tag{19}$$

2.2.2. Closure for Drag

The drag force $\mathbf{F}_D$ acting on a particle p is written:

$$\mathbf{F}_D = \frac{m_p}{\tau_p}(\mathbf{u}_{@p} - \mathbf{u}_p), \tag{20}$$

where $\mathbf{u}_{@p}$ is the local gas velocity interpolated at the center of the particle p and τ_p is the drag relaxation time expressed as follows:

$$\tau_p = \frac{4\rho_p d_p^2}{3\mu C_D Re_p} \quad \text{with} \quad Re_p = \frac{\varepsilon_{@p}\rho||\mathbf{u}_{@p} - \mathbf{u}_p||d_p}{\mu}, \tag{21}$$

where ρ_p is the particle density, d_p its diameter, μ the dynamic viscosity, $\varepsilon_{@p}$ is the local fluid fraction interpolated at the center of the particle, and C_D is the drag coefficient. In order to compute C_D, the classical closures proposed by Ergun ($C_{D,ER}$) [17] for high $\varepsilon_{@p}$ values and Wen & Yu ($C_{D,WY}$) [18] for low $\varepsilon_{@p}$ values are used along with the smoothing function ϕ_{gs} introduced by Huilin & Gidaspow [19] to avoid discontinuities when switching models:

$$\phi_{gs} = \frac{1}{\pi}\arctan(150 \times 1.75(0.2 - (1 - \varepsilon_{@p}))) + 0.5, \tag{22}$$

in such a way that:

$$C_D = \phi_{gs} C_{D,WY} + (1 - \phi_{gs}) C_{D,ER}, \tag{23}$$

with

$$C_{D,WY} = \begin{cases} \frac{24}{Re_p}\left(1 + 0.15 Re_p^{0.687}\right)\varepsilon_{@p}^{-2.7} & \text{for } Re_p < 10^3, \\ 0.44\varepsilon_{@p}^{-2.7} & \text{for } Re_p \geq 10^3, \end{cases} \tag{24}$$

and

$$C_{D,ER} = \left(200\frac{1 - \varepsilon_{@p}}{Re_p} + \frac{7}{3}\right)\varepsilon_{@p}^{-1}. \tag{25}$$

2.2.3. Other Forces

The gravity force $\mathbf{F}_G$ acting on a particle p is written:

$$\mathbf{F}_G = m_p \mathbf{g}. \tag{26}$$

The pressure gradient force $\mathbf{F}_P$ in Equation (5) reads as:

$$\mathbf{F}_P = -V_p \nabla P_{@p}, \tag{27}$$

where V_p is the particle's volume and $\nabla P_{@p}$ is the local pressure gradient interpolated at the center of the particle.

2.3. Phase Coupling

The coupling between the particle and fluid phases is a key point for the modeling of particle-laden flows, especially when the particle size approaches the Eulerian cell size. Many Eulerian fields have to be interpolated at the center of the particles for the numerous closures, as shown in Section 2.2. In the YALES2 solver, particles are located in a unique mesh cell (C) using the position of their center. For any Eulerian scalar or vector field $\Phi(\mathbf{x}, t)$, its value taken at the particle p center $\Phi_{@p}(t)$ obeys:

$$\Phi_{@p}(t) = \sum_{i \in C} \omega_{p,i} \Phi(\mathbf{x}, t) \quad \text{with} \quad \sum_{i \in C} \omega_{p,i} = 1. \tag{28}$$

Here i is a node index so that '$i \in C$' means 'all nodes composing the mesh cell C in which the particle p is located'. $\omega_{p,i}$ is the interpolation weight of the particle p on cell node i and is calculated using a trilinear interpolation on tetrahedra and on hexahedra. The same interpolation weights are used for data transfer from grid to particles (interpolation step) and from particle onto the grid (projection step).

The conservative projection operator needed to compute $\mathbf{F}_{p \to f}$ (see Equation (4)) is thus written on each node i as:

$$\mathbf{F}_{p \to f,i} = - \sum_{p \in SC_i} \omega_{p,i} (F_D + F_P), \tag{29}$$

and the fluid fraction at node i is written:

$$\varepsilon_i = \frac{1}{\Delta V_i} \sum_{p \in SC_i} \omega_{p,i} V_P. \tag{30}$$

Here ΔV_i denotes the control volume of node i and '$p \in SC_i$' means 'all particles belonging to any surrounding cell (SC) of node i'. Referring to Figure 1, it means that any particle belonging to one of the cells will be accounted for when computing ε (as well as $\mathbf{F}_{p \to f}$) at node 3.

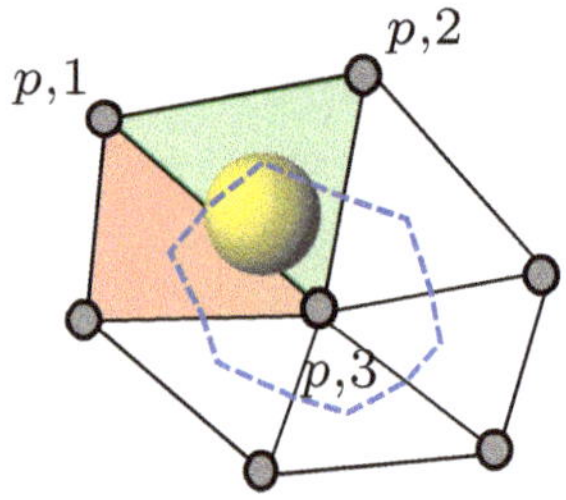

Figure 1. 2D representation of a particle p located in a triangular cell (green) moving towards a neighboring cell (red). ●: mesh nodes. - - -: contour of node 3 control volume. $\omega_{p,i}$: interpolation weight of particle p on node i.

This method consisting in distributing particle quantities only in the cell where its center resides, referred to as Particle Centroid Method (PCM), can lead to large calculation errors in particular regarding the fluid fraction, as pointed out in [20]. This is partly due to the fact that many CFD/DEM codes feature a staggered grid where the fluid fraction is defined at cell centers, causing dramatic discontinuities in time and space derivatives when a particle enters or leaves a cell. On the contrary, in the YALES2 code the fluid fraction, as all the Eulerian fields, is computed at the grid nodes. As depicted in Figure 1, it is then straightforward that the particle crossing from the green cell to the red one won't cause any discontinuity on the computation of neither $\omega_{p,1}$ nor $\omega_{p,3}$ involved in their interface. Moreover $\omega_{p,2}$ won't be much affected neither during the crossing because as the particle approaches the cell's interface, $\omega_{p,2}$ tends towards 0. This is still true in 3D cases and on Cartesian meshes.

Nevertheless, it is well known that the PCM method can induce inaccuracies and lead to numerical instabilities because it cannot prevent the fluid fraction from reaching unrealizable values, in particular when dealing with close to unity particle diameter/mesh cell size ratios. In such cases, the fluid fraction value can locally decrease below the theoretical packing limit. To cope with this limitation, a filtering operator well suited for distributed memory machines is used. Taking a 2D case as shown on the left in Figure 2, this filtering operator is built for any Eulerian scalar or vector field noted Φ_i, its filtered value being $\hat{\Phi}_i$. At node i_1, $\hat{\Phi}_i$ reads:

$$\hat{\Phi}_{i_1} = \frac{1}{3}\Phi_{i_1} + \frac{1}{3\Delta V_{i_1}}((S_{23}+S_{34})\Phi_{i_3} + (S_{34}+S_{45})\Phi_{i_4} + \cdots + (S_{72}+S_{23})\Phi_{i_2}), \tag{31}$$

where ΔV_{i_1} is the control volume associated to node i_1 and S_{mn} is the part of ΔV_{i_1} contained in the face delimited by nodes i_1, i_m and i_n, as shown on the left in Figure 2. If all the control volumes are equal, on a structured mesh for instance, Equation (31) becomes:

$$\hat{\Phi}_{i_1} = \frac{1}{3}\Phi_{i_1} + \frac{1}{9}\sum_{m\in[\![2;7]\!]}\Phi_{i_m}. \tag{32}$$

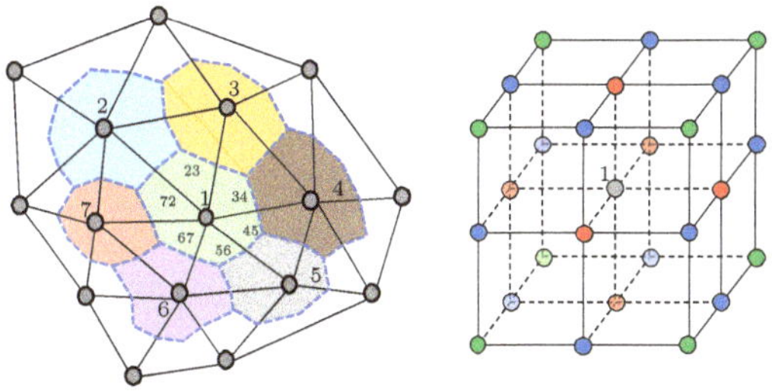

Figure 2. On the left: 2D representation of an unstructured mesh. ●: mesh nodes. The control volumes of nodes i_1 (■), i_2 (■), i_3 (■), i_4 (■), i_5 (■), i_6 (■), i_7 (■) are shown. The control volume of node i_1 is composed of surfaces S_{23}, S_{34}, S_{45}, S_{56}, S_{67} and S_{72}. On the right: 3D representation of a regular Cartesian mesh part. ●: node i_1. ●: nodes at faces' center. ●: nodes at edges' center. ●: nodes at corners.

The same type of filter can be derived in 3D. The following equation gives the value of $\hat{\Phi}_{i_1}$ in a 3D structured case with all equal control volumes as shown on the right in Figure 2, where RN is the abbreviation of Red Nodes, BN of Blue Nodes and GN of Green Nodes:

$$\hat{\Phi}_{i_1} = \frac{1}{8}\Phi_{i_1} + \frac{1}{64}\left(4\sum_{m\in RN}\Phi_{i_m} + 2\sum_{m\in BN}\Phi_{i_m} + \sum_{m\in GN}\Phi_{i_m}\right). \tag{33}$$

This fully conservative operation being performed on all volumes at the same instant provides a fully filtered field, and can be repeated several times to increase the filter width. In all the simulations presented in this paper, only one filtering step was used. It should be underlined that for the computation of $\hat{\varepsilon}$, the filtering step is applied before dividing by the local control volume in order to conserve total solid mass over the whole computational domain volume V:

$$\text{Total solid mass} = \rho_p \int_V (1-\varepsilon)dV = \rho_p \int_V (1-\hat{\varepsilon})dV. \tag{34}$$

The properties of such a filtering operator, i.e., its moments, are not straightforward to determine on unstructured meshes but it can be noticed that it is based on direct neighbors and thus doesn't need distant nodes, hence its attractiveness regarding parallelism. The main drawback is that the filter width can't be directly obtained because it depends on the local mesh size. Thus, when using this filtering operator, the user cannot prescribe the filter width. Other projection methods have been developed to circumvent this drawback while ensuring interesting mathematical properties as moment conservation for instance. One can cite the work of Capecelatro [8], who used a Gaussian kernel filter in a method called mollification for the same kind of application as presented here, and the work of Mendez [21] in which a projection operator based on high-order moments conservation was built for deformable red cell membrane modeling purposes.

2.4. Numerical Schemes

2.4.1. Fluid Advancement Procedure

This section presents some numerical features of the YALES2 code. This code solves the filtered low-mach number Navier-Stokes equations presented in Section 2.1 with an explicit time advancement. Among the various implemented numerical schemes, a fourth-order central scheme was used for the spatial integration, and a fourth-order scheme called TFV4A [22] combining Runge–Kutta and Lax-Wendroff methods was used for the explicit time integration.

The time advancement uses a time-staggered projection method for variable density flows [11] described below in which the n superscript refers to discrete times:

1. Lagrangian phase advancement
   First, the particles are advanced. The full description of this step is available in Section 2.4.2. After being relocated on the grid, $\varepsilon^{n+3/2}$ can be computed using Equation (30).
2. Density prediction for scalar advancement
   The density predictor $(\varepsilon\rho)^\star$ is then determined by the mean of the mass conservation equation (Equation (3)):
$$\frac{(\varepsilon\rho)^\star - (\varepsilon\rho)^{n+1/2}}{\Delta t} = -\nabla \cdot (\varepsilon\rho\mathbf{u})^n. \tag{35}$$
3. Velocity prediction
   Once $(\varepsilon\rho)^{n+1}$ is known, the velocity can be predicted reusing the dynamic pressure gradient of the previous time step:
$$\frac{(\varepsilon\rho\mathbf{u})^\star - (\varepsilon\rho\mathbf{u})^n}{\Delta t} = -\nabla \cdot ((\varepsilon\rho\mathbf{u})^n\mathbf{u}^n) - \nabla P_2^{n-1/2} + \nabla \cdot \boldsymbol{\tau}^n. \tag{36}$$
4. Velocity correction
   Velocity correction is performed by updating the pressure gradient:
$$\frac{(\varepsilon\rho\mathbf{u})^{n+1} - (\varepsilon\rho\mathbf{u})^\star}{\Delta t} = -\nabla(P_2^{n+1/2} - P_2^{n-1/2}). \tag{37}$$
   The Poisson equation aiming at calculating $P_2^{n+1/2}$ is obtained by taking the divergence of Equation (37) and inserting the condition imposed by the following equation of mass conservation written for $\mathbf{u}^{n+1}$:
$$\nabla \cdot (\varepsilon\rho\mathbf{u})^{n+1} = -\frac{(\varepsilon\rho)^{n+3/2} - (\varepsilon\rho)^{n+1/2}}{\Delta t}. \tag{38}$$
   The Poisson equation finally reads:
$$\nabla \cdot \left(\nabla \left(P_2^{n+1/2} - P_2^{n-1/2}\right)\right) = \frac{(\varepsilon\rho)^{n+3/2} - (\varepsilon\rho)^{n+1/2}}{\Delta t^2} + \frac{\nabla \cdot (\varepsilon\rho\mathbf{u})^\star}{\Delta t}. \tag{39}$$
   This linear system requires an efficient and accurate iterative solver. For all our simulations, a Deflated Preconditioned Conjugate Gradient (DPCG) algorithm [23] is used.

The resulting time advancement is fully mass and momentum conserving. The time step for the fluid phase Δt is calculated at each solver iteration by enforcing a maximum value of 0.2 for the Courant-Friedrichs-Lewy number (CFL) criterion and 0.15 for the Fourier number criterion for all the simulations.

2.4.2. Particle Advancement Procedure

A second-order explicit Runge–Kutta (RK2) algorithm is used to advance the position $\mathbf{x}_p$, the velocity $\mathbf{u}_p$ (see Equation (5)) of the particles in time:

$$\text{RK2 - 1st step:} \begin{cases} \mathbf{x}_p^{n+1/2} = \mathbf{x}_p^n + \dfrac{\Delta t}{2}\mathbf{u}_p^n, \\ \mathbf{u}_p^{n+1/2} = \mathbf{u}_p^n + \dfrac{\Delta t}{2}\dfrac{\sum \mathbf{F}^n}{m_p}, \end{cases} \tag{40}$$

$$\text{RK2 - 2nd step:} \begin{cases} \mathbf{x}_p^{n+1} = \mathbf{x}_p^n + \Delta t \mathbf{u}_p^{n+1/2}, \\ \mathbf{u}_p^{n+1} = \mathbf{u}_p^n + \Delta t \dfrac{\sum \mathbf{F}^{n+1/2}}{m_p}, \end{cases} \tag{41}$$

where $\sum \mathbf{F}$ refers to the summation of all forces acting on particle p (see Section 2.2).

At the particle scale, several phenomena need to be integrated properly by the mean of an associated characteristic time, among these: drag, gravity, etc. Thus, several stability criteria have to be computed on each particle to determine the smallest time step needed for the most constraining characteristic time. In dense fluidized beds simulations the collision time step is generally the limiting one, so it will be noted Δt_p from now on.

Δt_p must be inferior to the contact time T_C described in Equation (19) in order to be able to solve collisions properly. Furthermore, it must be small enough to ensure numerical stability depending on the selected numerical scheme, without compromising the performances of the code. It can be noted that there is a unique value of T_C because all the particles are identical. In our simulations, the following criterion was used:

$$\Delta t_p = T_C/6. \tag{42}$$

This criterion has been tested against the following non-dimensional analytical solution of Equation (16) that can be found in [24] for a normal collision of a single particle on a wall with given collision parameters as shown in Figure 3:

$$\delta_{ab}^{\star}(t) = \frac{\delta_{ab}(t)}{\delta_{ab}^{max}} = \frac{\omega_0}{\Omega} exp\left(\gamma_n \left[\frac{1}{\Omega} arcsin\left(\frac{\Omega}{\omega_0}\right) - t\right]\right) sin\left(\Omega t\right), \tag{43}$$

$$u_{ab,n}^{\star}(t) = \frac{u_{ab,n}(t)}{u_0} = \frac{1}{\Omega} e^{-\gamma_n t}\left(-\gamma_n sin(\Omega t) + \Omega cos(\Omega t)\right), \tag{44}$$

with

$$\Omega = \sqrt{\omega_0^2 - \gamma_n^2}. \tag{45}$$

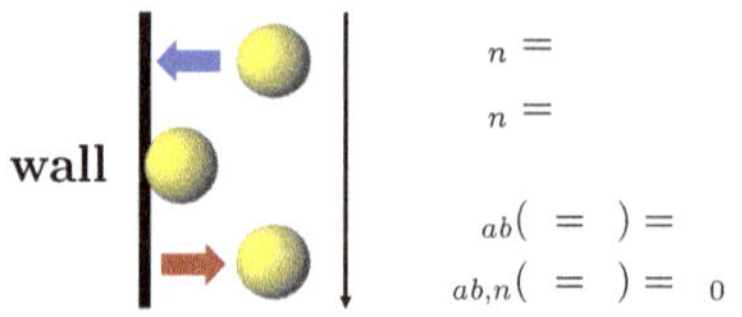

Figure 3. The simple test consists of a single particle colliding a wall in the normal direction. Only the collision force is accounted for and the parameters used for the resolution are visible on the right side. The contact starts at $t = 0$.

The time evolution of both overlap and particle velocity during contact has been plotted in Figure 4a,b, respectively.

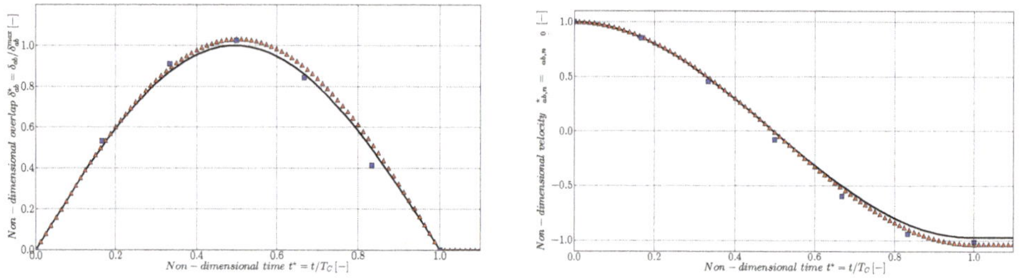

(**a**) (**b**)

Figure 4. Non-dimensional overlap (**a**) and velocity (**b**) as a function of the non-dimensional time during the contact presented in Figure 3. Comparison between analytical method (—), Euler method with $t_p = T_C/80$ (▲) and RK2 method with $t_p = T_C/6$ (■).

The agreement with the analytical solution is not perfect. However the scheme is stable and in this case the absolute error on the restitution velocity was found to lie below 5%.

2.5. *Performances and Parallelism*

A critical point in the coupling between the gas and solid phase is the time advancement, as they can have very different time scales. The choice was made to sub-step the solid phase, of which time advancement is generally limited by the collision time step in dense fluidized beds, during a gas phase time step, which is itself governed by convective and diffusive time scales. The mean number of particle sub-steps observed during our simulations was approximately 3 for the refined mesh, 7 for the intermediate mesh and 16 for the coarse mesh (mesh details are described in the following section).

As the computation of the collision force requires a distance estimation for each particle pair *a priori*, solving Equations (40) and (41) can become critically time consuming. The work of Lubachevsky [25] provides an analysis of the linked-cell method to tackle this problem, which is the most commonly used method. The solution consists in the definition of a Cartesian grid superimposed on the grid mesh allowing each particle to search for its potential collision partners, as shown in Figure 5. First, each particle is located in one of the Cartesian grid cell. Then looping over the 8 surrounding cells plus the one where the particle stands (26 + 1 cells in 3D) provides a list of the closest particles, that are stored as potential collision partners. Eventually, during the computation of the collision force, only these particles are checked.

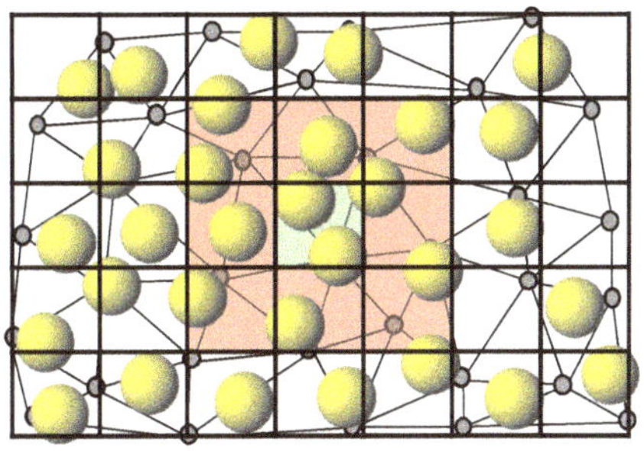

Figure 5. Linked-cell method. ●: mesh nodes. —: superimposed Cartesian grid. The particle of interest is noted p in the central cell (■). The particles of which center belongs to this cell or one of the surrounding ones (■) are stored as potential collision partners of particle p.

The linked-cell method is used even if the grid is Cartesian. This choice is justified by the fact that the cells of the detection grid should probably be (i) at least as big as the particle diameter (ii) as small as possible, in order to prevent too many neighboring particles from being detected. These requirements may be hard to guarantee if the Cartesian mesh is too coarse, or locally refined (which is one of the future aims of the code). For these reasons, a Cartesian grid is superimposed in any case.

Parallelism also requires special treatment for particles, as collision might occur between some of them although they don't belong to the same processor domain. To cope with this requirement, a ghost particle method is used, which is also classical in CFD/DEM. Ghost particles are identified using a cell halo surrounding each processor domain as shown in Figure 6 for processor ranked #1 in a cylindrical geometry discretized with an unstructured mesh. The particles belonging to the cell halo are exchanged between involved processors with the necessary data to compute collision force only by the mean of non-blocking Message Passing Interface (MPI) exchanges. The size of the cell halo is locally determined by the cell mesh size and particle diameter to avoid unwanted distant particles from being identified as ghost particles, which is not straightforward on unstructured meshes. On each processor, ghost particles are treated to be located as well on the afore mentioned Cartesian detection grid (see Figure 5).

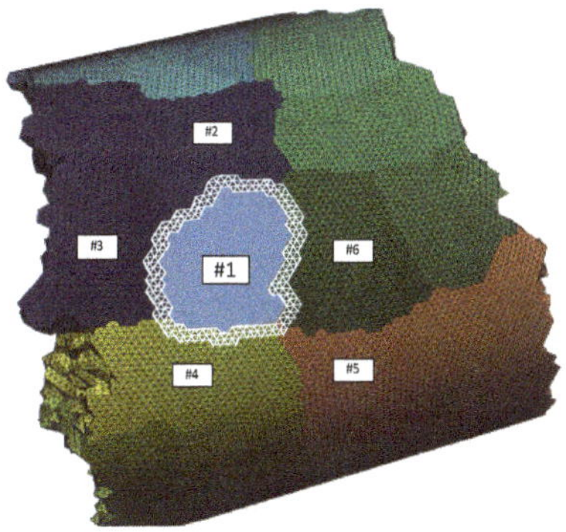

Figure 6. Ghost particle method principle shown for a part of a cylindrical domain. The different processor domains are colored accordingly. The processor of interest is ranked #1 (■) and its closest neighbors are ranked #2, #3, #4, #5 and #6. A particle entering the white cell halo around #1 will be sent to #1 as a ghost particle by the processor it belongs to.

2.6. Assessment of the Solver Performances

Assessments of the code global performances are first extracted from a canonical isothermal case, disregarding its physical relevance. This case consists of a cubic box meshed with tetrahedra. Particles are randomly seeded in the box, with a mean porosity of about 0.54, and each tetrahedron contains roughly 11 particles. A fluid phase is present to account for the computational cost related to interpolation and projection steps. The particle timestep is chosen such that, ten particle timesteps are performed for each fluid timestep, which corresponds to a usual sub-stepping configuration. Code assessments are obtained from a single fluid timestep. All the tests were carried out on the regional supercomputer Myria of the CRIANN center (Centre Régional Informatique et d'Applications Numériques de Normandie, France), featuring a Intel Omni-Path interconnect. The processors used are Intel Broadwell 14 cores running at 2.4 GHz with 128 GB RAM (about 4 GB memory per core) for total peak power of 400 TFlop/s.

The speed-up is first obtained by running the same simulation on different numbers of cores, ranging from 532 (reference case) to 4144. Each simulation roughly gathers 38M tetrahedra and 410M particles. The reference CPU time t_{CPU}^{ref} being associated with the temporal loop of the solver on $Nprocs^{ref} = 532$ cores, the speed-up for a CPU time t_{CPU} on $Nprocs$ is calculated by:

$$\text{speed-up}(Nprocs) = Nprocs^{ref} \times \frac{t_{CPU}^{ref}}{t_{CPU}}, \tag{46}$$

and is illustrated on Figure 7a. The solver exhibits a good scalability up to 4144 cores, with a speed-up reaching 80% of the ideal scaling.

Secondly, the scale-up of the solver is quantified by an evalutation of the performances at constant load per core on different numbers of cores, ranging from 252 (reference case) to 4144. The number of particles per core is about 99k, and the number of tetrahedra is about 9.1k per core. The scale-up is given by:

$$\text{scale-up}(Nprocs) = Nprocs^{ref} \times \frac{t_{CPU}^{ref}}{N_c^{ref}} \frac{N_c}{t_{CPU}}, \tag{47}$$

where N_c^{ref} and N_c are the number of cells in the reference case and in the current case, respectively. The scale-up curve is represented on Figure 7b, which shows an excellent scaling.

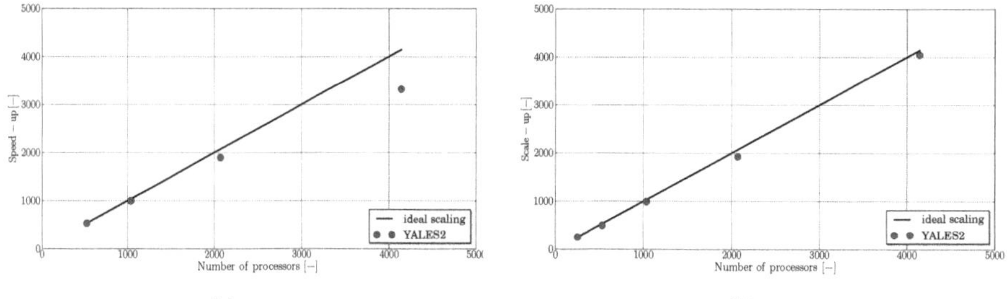

Figure 7. Speed-up (**a**) and scale-up (**b**) curves extracted from the canonical case.

Performances of the code are also assessed on an isothermal pressurized gas-fluidized bed carried out at the university of Birmingham [26]. The investigated case gathers 10M particles in a cylindrical domain meshed with 3.7M tetrahedra. Here, the performances are extracted over 1 s of physical time after the stability of the bed fluidization has been assessed by monitoring both the bed height and the pressure loss across the system. Thus, this case deals with realistic conditions where the execution speed highly depends on the local physics (presence of dense or void zones), and where, the fluid and particle timesteps are recomputed throughout the simulation. The tests were carried out on the Curie supercomputer of the TGCC center (Trés Grand Centre de Calcul, France), featuring a InfiniBand QDR Full Fat Tree interconnect. The used nodes comprise two Intel Sandy Bridge octo-core processors running at 2.7 GHz with 64 GB RAM (about 4 GB per core).

The speed-up is obtained by running the simulation on various numbers of cores, ranging from 64 (reference case) to 1024 cores, and is illustrated on Figure 8. On the whole, the solver reaches 55% of its ideal scaling value, which is acceptable given the high dispersion of the particles amongst the cores, causing their de-synchronizing.

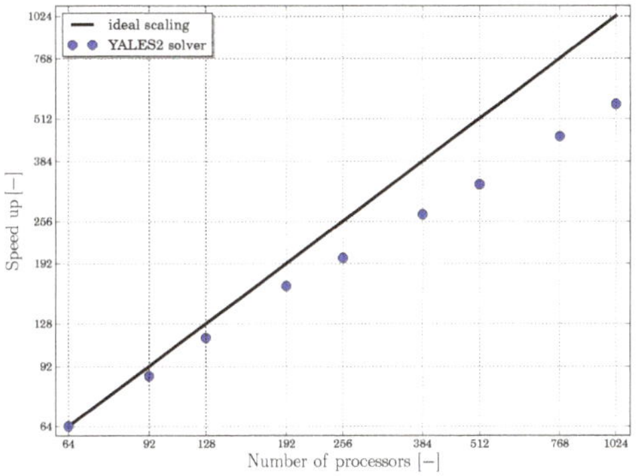

Figure 8. Speed-up curve obtained on the Birmingham fluidized bed.

3. Simulation Cases

3.1. *Configuration and Meshes*

In this work, CFD/DEM simulations of a fluidized bed, similar to that used in the experiments reported by Patil et al. [27], are performed. A sketch of the simulated configuration of the fluidized bed is shown in Figure 9 (left) including its dimensions. The bed contains inert glass particles fluidized by fresh nitrogen injected through the bottom of the bed at different gas flowrates. Physical properties of both gas and particles are summarized in Table 1. In the experiments, the bottom area of the bed was equipped with a small circular nozzle of 1.2 cm diameter, through which no gas was supplied during the experiments. In order to reproduce the effect of the switched nozzle as faithfully as possible, a zero gas flow was set through an area located at the center of the lower horizontal section of the bed. This area was considered as a wall for both gas and particles and is referred to as the bottom wall in the simulations. In the simulations, three different grid refinements were used and their characteristics are reported in Table 2, including that of the bottom wall retained for each mesh. As an example, a sketch of the intermediate mesh is illustrated in Figure 9 (right).

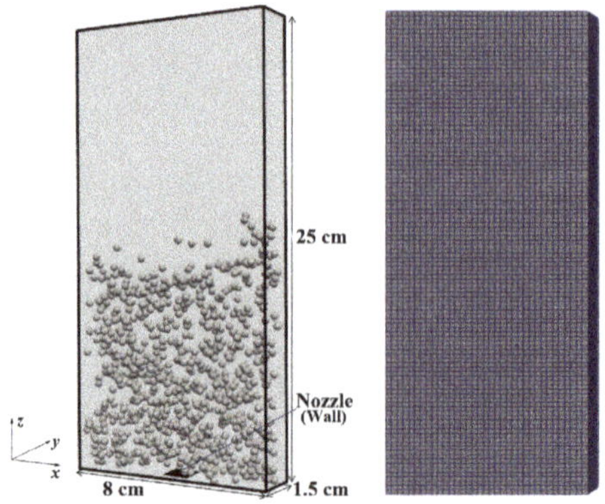

Figure 9. Sketches of the simulated configuration (**left**) and of the intermediate mesh (**right**).

Table 1. Gas and particle properties [27].

Gas Properties (at 20 °C)	
Density, ρ	1.165 kg/m^3
Viscosity, μ	2×10^{-5} Pa·s
Particle Properties	
Mean diameter, d_p	1 mm
Density, ρ_p	2500 kg/m^3
Norm. coef. of rest., e_n	0.97
Solid/gas density ratio, ρ_p/ρ	2146

Table 2. Mesh characteristics.

Mesh	Number of Cells				Bottom Wall		Mesh Size/Particle Diameter Ratio		
	N_x	N_y	N_z	Total	N_x	N_y	Δ_x/d_p	Δ_y/d_p	Δ_z/d_p
Coarse	17	3	55	2805	3	3	4.706	5.000	4.545
Intermediate	35	6	110	23,100	5	4	2.286	2.500	2.273
Fine	70	12	220	184,800	10	8	1.143	1.250	1.136

3.2. Description of the Simulation Cases

Several simulations based on different grid refinements, combined with different physical parameters, are investigated. In our modeling strategy, a friction coefficient, previously denoted as μ_{tan} in Section 2.2.1, is required to account for the tangential contact force during binary particle and particle-wall collisions. Since such a coefficient is not provided in the work of Patil et al. [27], we examined available values from the literature. As reported by Lorenz et al. [28], its value may depend on the state and history of the particle surfaces. The authors experimentally evaluated the friction coefficient, of glass particles having experienced a considerable time in granular flows, to 0.177 and 0.141 for inter-particle and particle-spent aluminum plate collisions, respectively. In the work of Goldschmidt et al. [29], values of 0.10 and 0.09, as measured by Gorham et al. [30], were respectively used in the modeling of the particle-particle and the particle-wall frictional collisions in a dense gas-fluidized bed configuration. In Gorham et al., both the particles and the pseudo-2D bed were made of glass. In the following sections, we denote the particle-particle and the particle-wall friction coefficients as μ_p and μ_w, respectively. All the simulations are summarized in Table 3. A mesh sensitivity analysis, drawn from simulations C1, C2 and C3, for which $\mu_p = \mu_w = 0.2$ is assumed, is first presented. In a second step, various values of the friction coefficients, μ_p and μ_w, are investigated.

This study refers to the sets of simulations C4 and C5, as shown in Table 3. The C4 simulations are performed using a constant value of μ_p, with different values of μ_w, whereas in the C5 simulations, the effect of μ_p is investigated for a constant value of μ_w. From these investigations, values for the friction coefficients are chosen by comparing numerical results of the volume fraction and the axial flux of the particulate phase with the available experimental measurements. Finally, additional Reynolds number and bed weight conditions are used as variable parameters for the simulations C6 and C7. It has to be mentioned that the inlet gas velocity, U_f, given in Table 3 is the superficial velocity through the horizontal cross-sectional area of the bed (so, this area also includes the bottom wall). For each tested gas velocity, the corresponding particle Reynolds number, given as $Re_p = \rho U_f d_p/\mu$, is reported in the table, together with the Stokes number $St = \rho d_p U_f/(18\mu)$. It has to be noted that, the bulk Reynolds number based on the bed width is relatively high ($5600 \leq Re \leq 8000$). Therefore, for all the performed simulations, a turbulent viscosity based on Smagorinsky model is used. The additional parameter of interest is the restitution coefficient during particle-wall collisions, e_w, which is considered equal to that of the particle-particle collisions, e_n.

Table 3. Simulation cases with the selected physical and numerical parameters.

Simulation Cases	Bed Mass (g)	U_f (m/s) at 20 °C	Re_p	St	Friction Coefficients μ_p	μ_w	Mesh
C1							Coarse
C2	75	1.20	70	~4	$\mu_p = \mu_w = 0.2$		Intermediate
C3							Fine
						0.0	
						0.2	
C4	75	1.20	70	~4	0.1	0.2	Intermediate
						0.4	
						0.8	
					0.0		
					0.2		
C5	75	1.20	70	~4	0.2	0.1	Intermediate
					0.4		
					0.8		
C6	75	1.71	100	~5.5	$\mu_p = \mu_w = 0.1$		Intermediate
C7	125	1.54	90	~ 5	$\mu_p = \mu_w = 0.1$		Intermediate

4. Results

Non-dimensional results are presented in this section. Since the lengths of the simulated configuration are different in all three directions a different length scale is defined in each direction. As a consequence, $x-$, $y-$ and $z-$distances are divided by the bed width ($L = 0.08$ m), the bed depth ($D = 0.015$ m) and the bed height ($H = 0.25$ m), respectively. The time scale is defined as $t/(d_p/U_f)$. The velocities are divided by the inlet gas velocity (U_f) and the mass flowrates by the inlet gas mass flowrate (ρU_f). In order to characterize the bed hydrodynamics in an Eulerian fashion, the instantaneous fields of any Lagrangian variable, ψ_p, are obtained by performing a spatial average over the grid cell. In addition, a 2D-behavior is considered for this fluidized bed configuration since the hydrodynamic variations in the depth direction are negligible. This can be observed in Figure 10, which displays the instantaneous particle volume fraction, α_p, on different slices chosen at $y/D = 0.2$ (close to the back side), $y/D = 0.5$ (in the middle of the bed) and at $y/D = 0.8$ (close to the front side).

As an example, in Figure 11, profiles along the depth direction at the height $z/H = 0.16$ are also given for the time-averaged vertical velocity of the solid phase normalized by the inlet gas velocity ($\overline{U_{p,3}}/U_f$). It can be observed that variations along the depth direction of such a quantity are not strong. Thus, two-dimensional variables are computed by averaging over the bed thickness (i.e., in the y-direction). In the present work, the 2D-variables corresponding to the computed field, $< \psi_p >$, are denoted as $< \psi_p >_{xz}$. All the simulations are run for 20 s in order to reach a steady regime. In such a regime, particles and fluid hydrodynamic fields are statistically stationary and time-averaged quantities, $\overline{< \psi_k >}$, for which k denotes the gas (g) or the particulate (p) phase, may be computed for comparison with the experimental measurements.

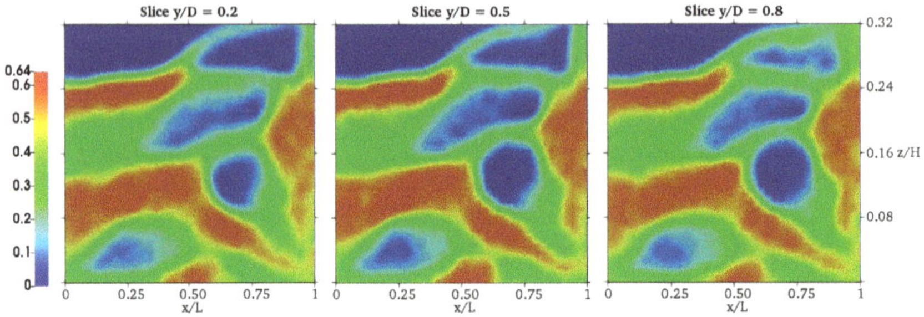

Figure 10. Instantaneous particle volume fraction on different slices of the bed. Simulation C2, dimensionless time $t/(d_p/U_f) = 12{,}000$, $\mu_w = \mu_p = 0.2$, $Re_p = 70$ and bed weight 75 g.

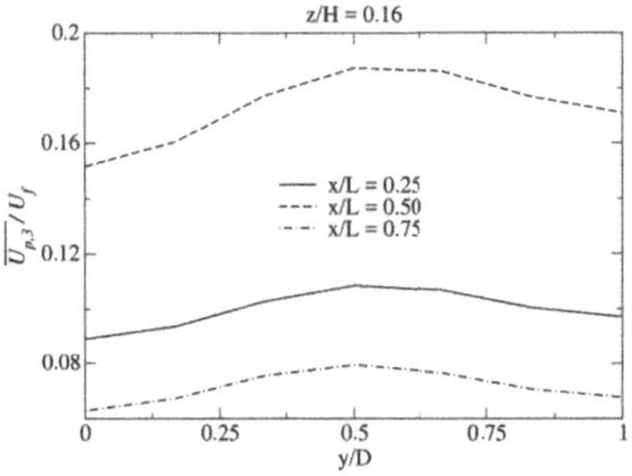

Figure 11. Profiles of the time-averaged normalized vertical velocity of the particulate phase ($\overline{U_{p,3}}/U_f$) at different locations x/L. Simulation C2, $\mu_w = \mu_p = 0.2$, $Re_p = 70$ and bed weight 75 g.

4.1. *Effect of the Grid Refinement*

In this section, the simulations C1, C2 and C3 are presented and discussed. The three simulations differ from each other by their mesh size, as shown in Tables 2 and 3, and consequently by their filtering kernel length ($\simeq 4\Delta x$). However, because the kernel length is always larger than the particle diameter, the volume-averaged equations are well posed in the sense of Jackson [7]. Snapshots of the time-averaged fields concerning the volume fraction, $\overline{< \alpha_p >_{xz}}$, and the mass flux, $\overline{< \alpha_p \rho_p U_p >_{xz}}$, of the particulate phase are depicted in Figure 12 for the different grid refinements. Experimental results are also included in the right column. The simulations match quite well the experiments. The particle volume fraction exhibits high values (dense regions) close to the side and bottom walls and small values (dilute regions) in the center of the bed. However, the average height of the fluidized bed seems to be slightly overestimated by the simulations. The solid flux exhibits a macroscopic double mixing loop and, on average, the particles move upward at the center of the bed and downward close to the side walls. Comparison to experiments reveals that the center of the mixing loops, as predicted by the simulations, is located slightly higher than that experimentally observed. The three simulations indicate clearly that the mesh size affects the bed hydrodynamics prediction. The reason is that particles continuously interact with the gas flow through the drag and the pressure gradient forces, which then contributes to the particle dynamics evolution. Since the mesh resolution affects the hydrodynamics prediction of the gas phase, this consequently modifies the particle dynamics predictions. The upper panel of Figure 12, shows that the finest grid leads to the best predictions, as a result of an accurate resolution of the gas hydrodynamics. The high solid concentrations close to the side walls, as reported by the experiments, are reproduced with a relatively good accordance by the fine-grid simulation. Nevertheless, refining the mesh does not seem to have a significant effect on the height of these dense regions from the bottom of the bed, nor on the mean bed height. Indeed, refining the mesh "locally" improves the bed hydrodynamic predictions by improving the fluid velocity field around the solid particles, but it does not influence the predicted overall bed height. This latter is instead directly affected by the selected drag force model. This is the main reason why one-dimensional models are accurate regarding the prediction of the mean bed height, given a sufficiently appropriate drag model. The analysis of the time-averaged fields of the solid flux reveals, apart from the progressive intensification of these fields due to the increasing mesh resolution, that this latter did not influence the position of the observed double mixing loop.

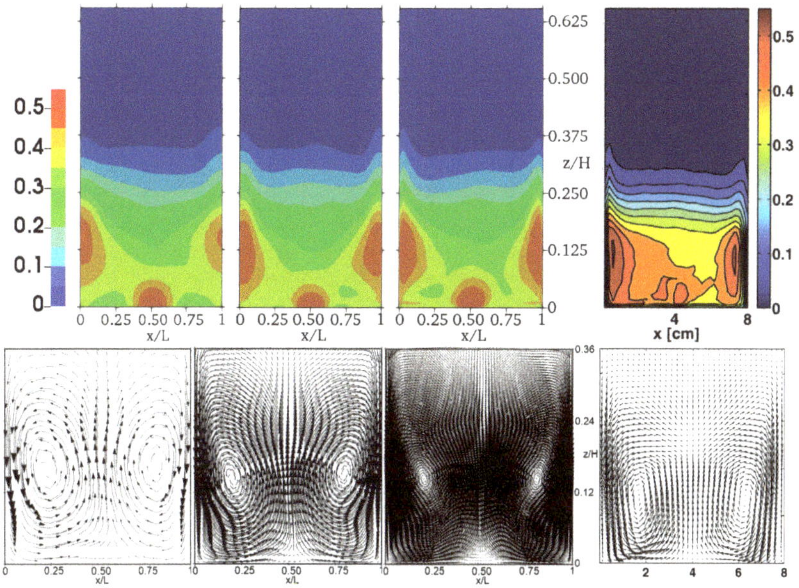

Figure 12. Numerical predictions of the time-averaged fields of the volume fraction, $\overline{<\alpha_p>_{xz}}$, (upper panel) and the mass flux, $\overline{<\alpha_p \rho_p U_p>_{xz}}$, (bottom panel) of the solid phase. From left to right, increasing grid refinement with $\mu_w = \mu_p = 0.2$, and comparison with the experiments (right column). $Re_p = 70$ and bed weight 75 g.

Figure 13 shows profiles of the time-averaged vertical solid mass flux normalized by the gas mass flowrate, $\overline{<\alpha_p \rho_p U_{p,3}>_{xz}}/\rho U_f$. The profiles are taken at the height $z/H = 0.092$ ($z = 2.3$ cm) above the bottom of the bed, for which experimental data are available. We notice that the predictions are improved when the finest mesh is used, especially close to the side walls. However, at the center of the bed, the effect of the bottom wall on the axial solid flux is still poorly reproduced even when a finer mesh is used.

With the attempt to improve the numerical results, the wall boundary condition type on the gas phase was also investigated. Within the framework of Navier-Stokes equations, the wall boundary condition for the gas is no-slip. However, such a condition is questionable for averaged equations, as those employed in this work. In addition, the presence of the particles may profoundly modify the flow hydrodynamics. In the present work, slip and no-slip fine-grid simulation cases were run and compared to each other. The time-averaged results (not shown here) relative to the particulate phase were only very slightly affected by the selected wall condition of the gas phase. Such a behavior is inherent to inertial particles, as those simulated in the present work, for which the Stokes number is much greater than unity, as shown in Table 3.

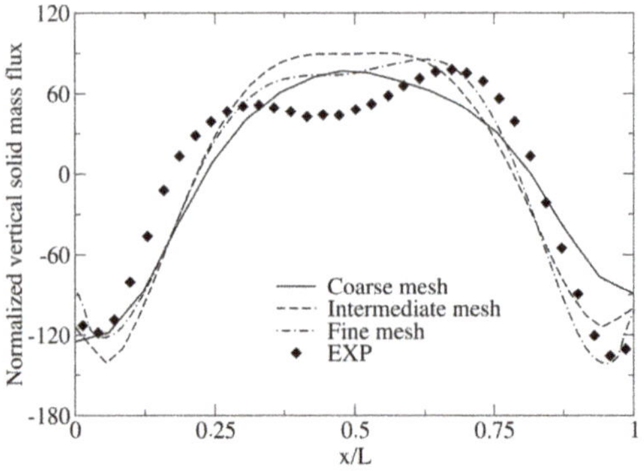

Figure 13. Time-averaged normalized vertical mass flux of the solid phase, $\overline{<\alpha_p \rho_p U_{p,3}>_{xz}}/\rho U_f$, at the height $z/H = 0.092$ above the bottom bed, for different grid refinements. $\mu_w = \mu_p = 0.2$, $Re_p = 70$ and bed weight 75 g.

4.2. *Investigation of the Frictional Effects*

Effects of the inter-particle and the particle-wall friction coefficients on the bed hydrodynamics are here discussed. In the previous study, the computational time required by the fine-grid simulation was nearly twice higher than that of the intermediate-grid simulation. Therefore, the present study is carried out using the intermediate mesh.

The time-averaged fields concerning the particle volume fraction, as obtained from the sets of simulations C4 and C5 are shown in Figure 14. As previously mentioned, a dense region is observed close to the side and bottom walls, whereas, the center of the bed is characterized by a dilute solid concentration. In addition, the average height of the fluidized bed seems to decrease as the frictional effects become stronger. However, more subtle differences between the situations can be observed. The upper panel as obtained by the simulations C4, reveals that, for the same value of the particle friction coefficient, when the wall frictional effects become stronger, the dense regions become denser and larger near the side walls, while they are sligthly narrowing above the bottom wall. Results also show that such regions move slightly upwards as μ_w increases. It has to be reminded that the influence of the wall friction coefficient is also exerted at the front and the back walls, but should be stronger in the confined parts of the bed, namely at the side walls, where the denser regions of the bed are observed. The bottom panel, obtained by the simulations C5, shows that, when μ_p increases up to 0.2, the bed hydrodynamics exhibit similar behavior as that observed when μ_w varies and μ_p is kept constant (simulations C4). Such results are in agreement with the work of Yang et al. [31], who showed the same trend of particle volume fraction distribution when μ_p was increased from 0.05 to 0.15, in their comparison study between two fluid model (TFM) and discrete particle model (DPM) simulations. However, unlike simulations C4, continuing to increase the particle-particle friction coefficient above the value of 0.2 in the simulations C5 resulted in decreasing the solid concentration near the side walls and increasing it above the bottom wall. The inter-particle friction acts on the whole bed volume and, as a consequence, high values of the inter-particle friction coefficient may considerably modify the bulk-bed hydrodynamic behavior. Finally, when the particle friction is not considered ($\mu_p = 0$ for $\mu_w = 0.1$), a homogeneous fluidized bed is obtained, which is not the case for $\mu_p \neq 0$ and $\mu_w = 0$ (simulations C4). These various bed behaviors inform about the effect of the particle friction on the bubble formation. As an example, snapshots of the instantaneous bed porosity (gas volume fraction), obtained at different instants by the simulations C5 for two values of μ_p, are depicted in Figure 15. In the case of frictionless particles ($\mu_p = 0$), no bubble formation occurs, whereas in the case of relatively frictional particles ($\mu_p = 0.2$), a realistic bubble growth is observed. This is consistent with the work of Hoomans et al. [32], who showed the same bubble patterns in 2D-fluidized bed DPM simulations.

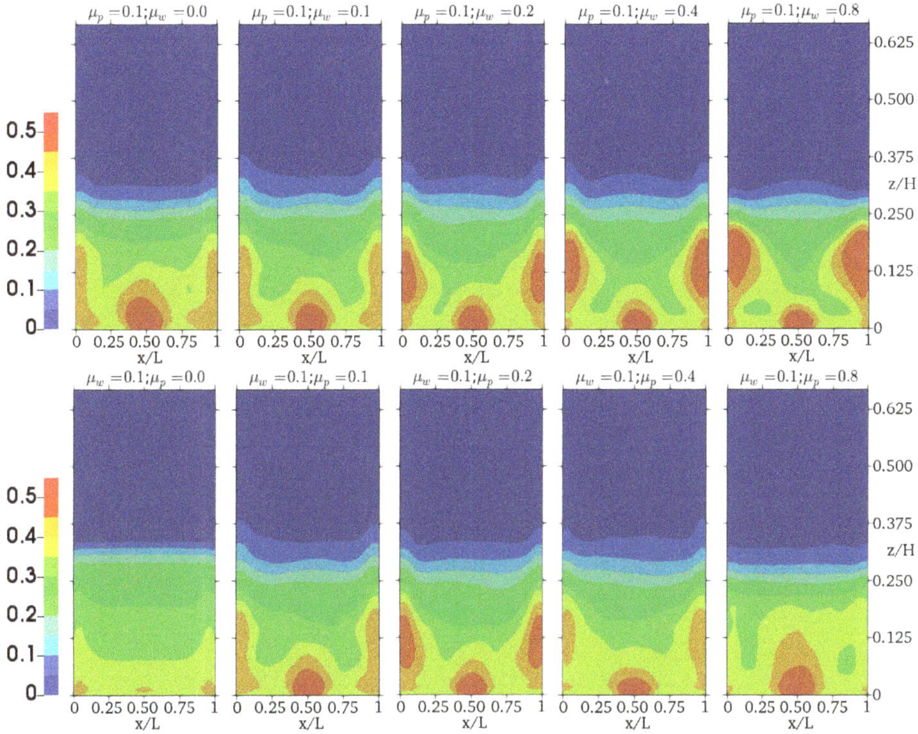

Figure 14. Time-averaged particle volume fraction, $\overline{<\alpha_p>_{xz}}$, for different values of the particle–wall μ_w and the particle–particle μ_p friction coefficients. Upper panel: constant μ_p and different μ_w. Bottom panel: constant μ_w and different μ_p. $Re_p = 70$ and bed weight 75 g.

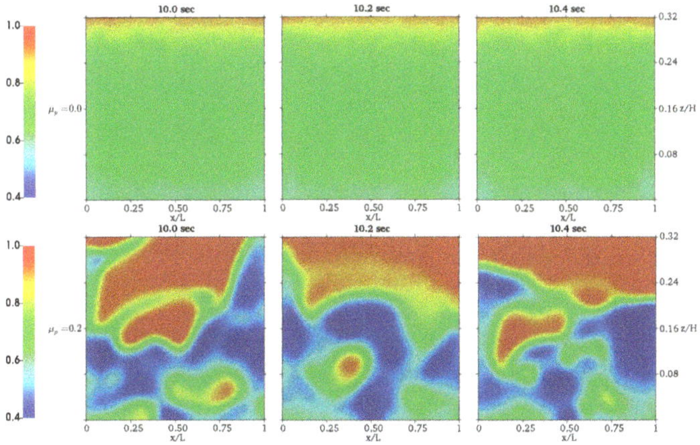

Figure 15. Instantaneous gas volume fraction, with $\mu_w = 0.1$. Upper panel: $\mu_p = 0$. Bottom panel: $\mu_p = 0.2$. $Re_p = 70$ and bed weight 75 g.

Time-averaged particle agitation energy, q_p^2, has been computed as $q_p^2 = \overline{u'_{p,i}u'_{p,i}}$, where $u'_{p,i} = u_{p,i} - U_{p,i}$. In these expressions, $u'_{p,i}$ and $U_{p,i}$ are the ith components of the fluctuating and the mean particle velocities, respectively, with the mean particle velocity computed as $U_{p,i} =< u_{p,i} >$. The normalized results are shown in Figure 16 for the two sets of simulations. The first panel obtained from the set C4 shows that most of the particle agitation is produced at the approximate height $z/h = 0.25$. This height corresponds to the average height of the dense bed ($\alpha_p > 0.2$). Furthermore, it increases when friction at wall becomes stronger. Results of the simulations C5, as depicted in the bottom panel, show a decline, on average, of the particle agitation energy when the particle friction increases. In addition, some particle agitation is produced at the height $z/h = 0.25$, similarly to what is observed in the C4 simulations. On the contrary, fewer amount of particle agitation is produced as the particle frictional effects become more significant. These results show that, for low friction between particles, the fluidized bed may be considerably heterogeneous when friction at wall is high and become homogeneous when friction between particles is strong. It has to be noted that, in all these simulations, the computed particle agitation energy in the upper part of the bed ($z/H \geq 0.375$), has a very low significance since, in this part, the particle volume fraction is nearly zero.

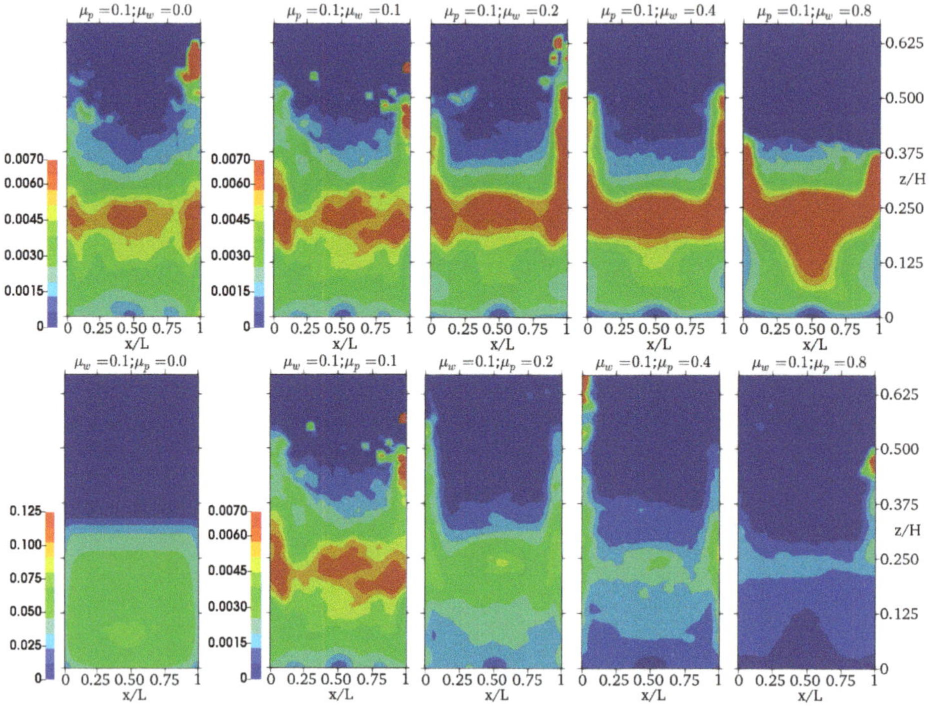

Figure 16. Time-averaged dimensionless particle agitation energy, q_p^2/U_f^2 for different values of the particle-wall μ_w and the particle-particle μ_p friction coefficients. Upper panel: constant μ_p and different μ_w. Bottom panel: constant μ_w and different μ_p. $Re_p = 70$ and bed weight 75 g.

Figure 17 shows the time-averaged field of the solid mass flux for the two sets of simulations. From the visualizations, it appears that friction affects significantly the magnitude of the solid flux and the extent of the mixing loops. In some cases, it also affects their position from the bottom of the bed. As previously observed for the particle volume fraction distribution, the position of the centre of the double mixing loop moves upwards when μ_w increases (upper pannel). This point is not straightforward when μ_p is increased (bottom panel). However, qualitatively speaking, the magnitude of the solid flux exhibits a maximum for $\mu_p = 0.2$ in both sets of simulations.

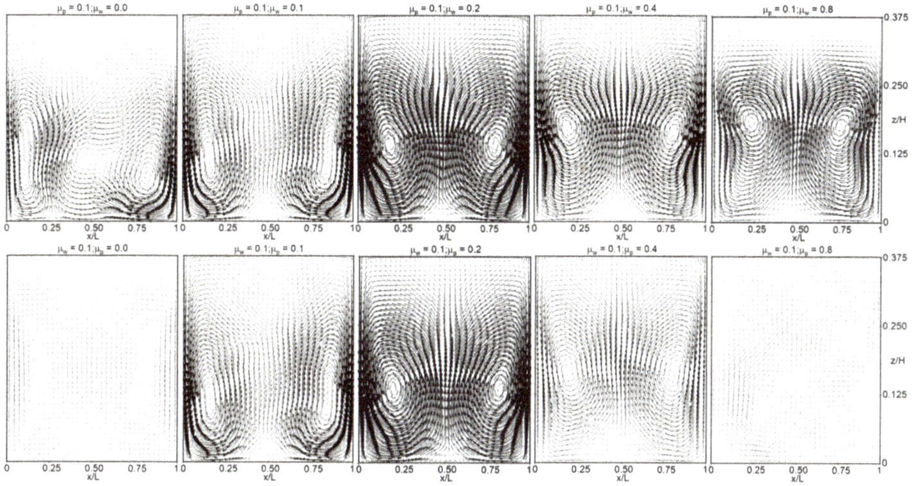

Figure 17. Time-averaged solid mass flux, $\overline{< \alpha_p \rho_p U_p >_{xz}}$, for different values of the particle–wall μ_w and the particle–particle μ_p friction coefficients. Upper panel: constant μ_p and different μ_w. Bottom panel: constant μ_w and different μ_p. $Re_p = 70$ and bed weight 75 g.

A quantitative analysis of the solid flux may be performed via its vertical profiles at a given height of the bed. Profiles of the time-averaged normalized vertical solid mass flux are depicted in Figure 18a,b for the two sets of simulations. The profiles are taken at the height $z/H = 0.092$ ($z = 2.3$ cm) above the bottom of the bed, for which experimental data are available. In all the simulations, two distinct parts for each profile can be observed. The first part corresponds to a nearly flat profile located in the region spanning the center of the bed and the second one is a sloped profile starting from the side walls. The expansion of one part to the detriment of the other depends on the values of the friction coefficients μ_w and μ_p. These coefficients also determine the maximum values of the upward and the downward mass fluxes in the flat and the sloped parts, respectively. The figures show that when μ_w or μ_p increases up to 0.2, the downwarding flux increases and the upwarding flux, close to the center of the bed, decreases. This leads, at the same time, to a steeper slope of the solid mass flux profile in the sloped part and to a less broad flat part at the center of the bed. For a stronger friction, either at wall or between particles, larger sloped parts are still observed, but with less steep slopes, leading to the decrease of both the downwarding and the upwarding solid fluxes. Therefore, despite the differences observed concerning the volume fraction and the agitation energy of the particulate phase, simulations C4 and C5 predict similar characteristics (field and magnitude) of the solid mass flux. It is conjectured that, in this confined configuration, when the wall friction coefficient increases, the front and back walls may considerably contribute together with the side walls to propagate stronger wall frictional effects towards the bulk bed. This has already been shown in a previous CFD/DEM study of Li et al. [33] for bed thicknesses of 10 and 20 particle diameters in a bubbling fluidized bed. As a consequence, stronger wall friction may produce almost the same effect on the solid flux as that induced by the friction between particles.

Finally, comparison with the experimental measurements shows that, the case with $\mu_p = \mu_w = 0.1$, is particularly interesting as it allows to reproduce the effect of the bottom wall on the axial solid flux (Figure 18) and to better predict the location of the solid mixing loops from the bottom of the bed (Figure 17 vs. experimental data from Figure 12). Nevertheless, the solid concentration close to the side walls seems to be slightly underestimated (Figure 14 vs. experimental data from Figure 12) and the solid loops more elongated than that experimentally reported. However, the predicted value of 0.1 for both the particle-particle and the particle-wall friction coefficients is noteworthy. In fact, the friction coefficient is a physical parameter that depends on the mechanical properties and the surface morphology of the solid particles, mainly Young's modulus and the surface roughness. Its experimental measurement is also very sensitive to some environmental factors, such as wet or dry conditions and the exerted normal stress. For this reason, a wide range of values are reported in the literature for glass beads under various conditions. Values ranging from 0.139 to 0.464 have been determined by Ishibashi et al. [34] for 1.6 mm-sized particles under dry conditions and various normal forces. Sandeep and Senetakis [35] reported a unique value of 0.2 for glass particles of 2 mm diameter. Additionally, other values may also be found in the literature as this is previously discussed in Section 3.2. The predicted value of 0.1 in our simulations is deemed to be acceptable and it is in very good agreement with that used by Goldschmidt et al. [29] in their simulation of glass particles in a pseudo-2D dense fluidized bed.

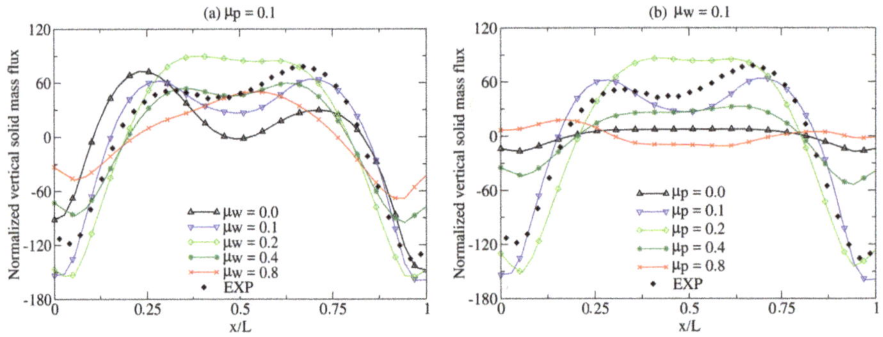

Figure 18. Time-averaged vertical solid mass flux normalized by the gas mass flow rate, $\overline{<\alpha_p \rho_p U_{p,3}>_{xz}}/\rho U_f$, for different values of the particle–wall and particle–particle friction coefficients, and comparison with experiments. $Re_p = 70$ and bed weight 75 g.

4.3. Application of the Results to Other Physical Configurations

In order to generalize the previous results to different physical configurations, other Reynolds number and bed weight conditions are investigated using the intermediate mesh. The same value of μ_w and μ_p and equal to 0.1 is used in the simulations since, as shown in Section 4.2, this leads to minor deviations from the experiments conducted for $Re_p = 70$. A gas no-slip condition at the wall is also retained. In the simulation C6, a superficial gas velocity of 1.71 m/s corresponding to $Re_p = 100$, as reported in Table 3, is imposed at the bottom of the bed. Predictions obtained by this simulation concerning the time-averaged fields of volume fraction and mass flux of the solid phase are reported in Figure 19a,b, respectively. Experimental measurements are also included for comparison purpose. One can see that the simulation reproduces very well the experimental solid distribution and the solid mass flux field. However, slight mismatches can be observed concerning the mean bed height and the position of the solid mixing loops. At the height $z/H = 0.092$, numerical and experimental vertical mass fluxes are displayed on Figure 20. Globally, results are conveniently reproduced by the simulation and confirm the accuracy of the selection $\mu_w = \mu_p = 0.1$.

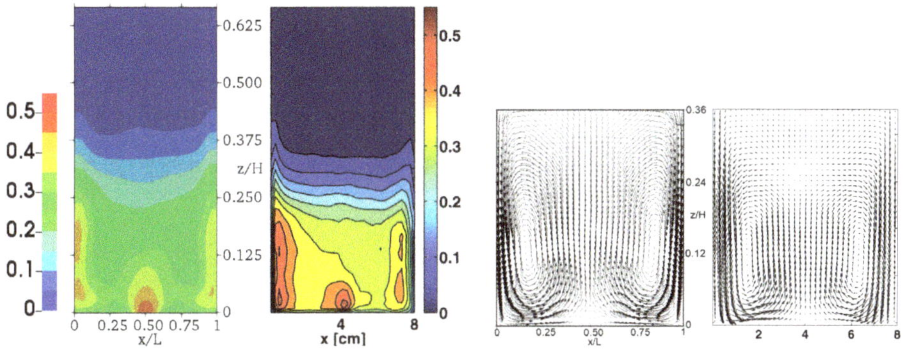

(**a**) $\overline{<\alpha_p>_{xz}}$ (**b**) $\overline{<\alpha_p\rho_p U_p>_{xz}}$

Figure 19. Numerical (**a**) against experimental (**b**) time-averaged fields of the volume fraction and the mass flux of the solid phase. $Re_p = 100$ and bed weight 75 g. In the simulation, $\mu_w = \mu_p = 0.1$.

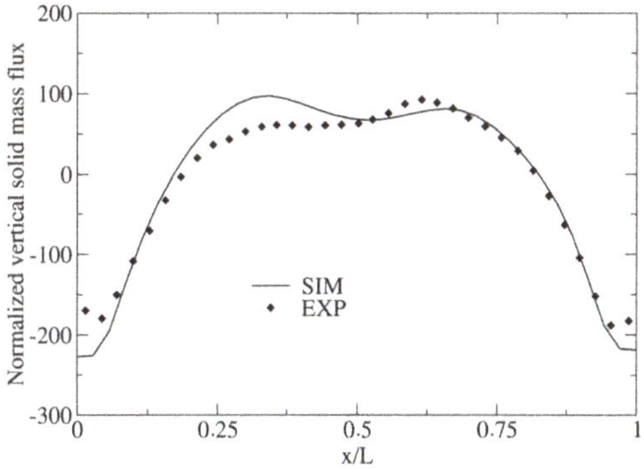

Figure 20. Numerical against experimental time-averaged vertical mass flux, $\overline{<\alpha_p\rho_p U_{p,3}>_{xz}}/\rho U_f$, of the solid at the height $z/H = 0.092$ above the bottom of the bed. $Re_p = 100$ and bed weight 75 g. In the simulation, $\mu_w = \mu_p = 0.1$.

Results of a fluidization experiment of 125 g of particles by nitogen injected at a velocity of 1.54 m/s, corresponding to $Re_p = 90$, have also been reported in Patil et al. [27]. This operating point is simulated in the present work (simulation C7) and results displayed in Figure 21a,b, together with the experimental data. Based on the investigations reported in Section 4.2, a value of the friction coefficient that allows to match as best as possible the experiments is retained. However, for all the simulations investigated so far, common deviations from the experiments are observed, namely concerning the mean bed height that appears overestimated, and the lateral solid recirculation zones which seem to be more elongated than that experimentally reported. The present simulation, using a higher bed weight, also does not sufficiently match the experiments, although the general pattern of the bed is globally well reproduced. Improvement of the numerical results should be achieved by, first, selecting an accurate drag law, which would allow to decrease the mean bed height, and then, by refining the mesh, which, as shown in Section 4.1, allows to accurately reproduce the solid concentration near the side walls.

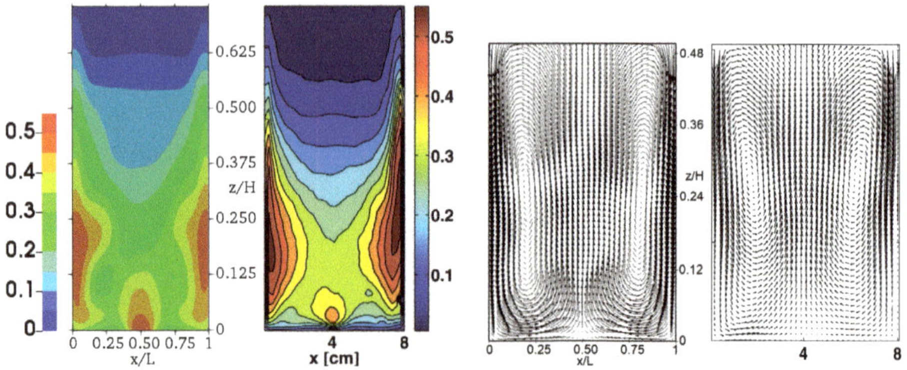

(**a**) $\overline{<\alpha_p>_{xz}}$ (**b**) $\overline{<\alpha_p \rho_p U_p>_{xz}}$

Figure 21. Numerical (**a**) againt experimental (**b**) time-averaged fields of the volume fraction and the mass flux of the solid phase. $Re_p = 90$ and bed weight 125 g. In the simulation, $\mu_w = \mu_p = 0.1$.

5. Conclusions

In this work, a DEM/LES approach is used to simulate a confined fluidized bed configuration for several physical and numerical conditions. Results are compared with experimental data of [27]. This study shows that, compared with the experimental measurements, predictions regarding the volume fraction and the mass flux of the particulate phase, obtained from the fine grid, are slightly better than that of the coarse and the intermediate grids. This slight improvement is attributed to a better prediction of the gas-particle interactions through the drag and the pressure gradient forces. The present work is also dedicated to frictional effects between particles and between particles and walls on the bed hydrodynamics behavior. It is shown that, increasing either the coefficient of the inter-particle friction or that of the particle-wall friction, the average bed height decreases and the bubble formation is enhanced. At low friction between particles, increasing the wall friction coefficient leads to similar solid circulation patterns (fields and magnitudes) as in the simulations with increasing friction between particles. Thus, it is conjectured that, in such confined 2D-configuration, when the wall friction coefficient increases, the friction of the front and back walls may have a significant impact on the inner-flow hydrodynamics of the fluidized bed. This finally may lead to the same bed hydrodynamics behavior as that observed when the friction coefficient between particles increases. Nevertheless, this point has to be further investigated. Indeed, to clearly understand the contribution of the friction at the front and back walls, simulations with various bed thicknesses combined with different values of the friction coefficients have to be performed. Additional simulations, performed at higher Reynolds numbers and/or other bed weights, showed globally good agreement with the experiments, reproducing almost similar particle volume fraction and solid circulation patterns as the experiments. However, deviations from the experiments concerning the average height of the dense bed is still exhibited in all the simulations. This could be further enhanced by studying other modeling features, such as the drag law.

Author Contributions: Investigation, Z.H.; Methodology, Y.D.; Supervision, J.-L.P., R.B., G.L. and V.M.

Funding: Computations were performed on the IFPEN supercomputer ENER110 and on the supercomputer CURIE under the project GENCI A0032B07345. The ENER110 and GENCI supercomputers teams are gratefully acknowledged. This work is done in the frame of the MORE4LESS project funded by the French National Research Agency.

Acknowledgments: Olivier Simonin at Institut National Polytechnique de Toulouse (France) is gratefully acknowledged for fruitful discussions throughout this work.

Conflicts of Interest: The authors declare no conflict of interest. The funders had no role in the design of the study; in the collection, analyses, or interpretation of data; in the writing of the manuscript, or in the decision to publish the results.

Abbreviations

The following abbreviations are used in this manuscript:

CFL	Courant-Friedrichs-Lewy
DEM	Discrete Element Method
DPCG	Deflated Preconditioned Conjugate Gradient
DPM	Discrete Particle Model
DNS	Direct Numerical Simulation
LES	Large-Eddy Simulation
MPI	Message Passing Interface
PCM	Particle Centroid Method
RK	Runge–Kutta
SC	Surrounding Cell
TFM	Two Fluid Model
BN	Blue Nodes
GN	Green Nodes
RN	Red Nodes

Appendix A. Influence of LES Model on the Numerical Results

In order to clearly assess the influence of the LES model on the numerical simulations we performed additional simulations with and without LES model. The physical and numerical parameters corresponds to case C6 in Table 3. Figure A1 shows that there almost no difference between the case using a LES model and the case with no model. The closure law for small scale structures have a weak impact on the flow which means that the dissipation is mainly governed by the large scale structure and by particle-particle friction.

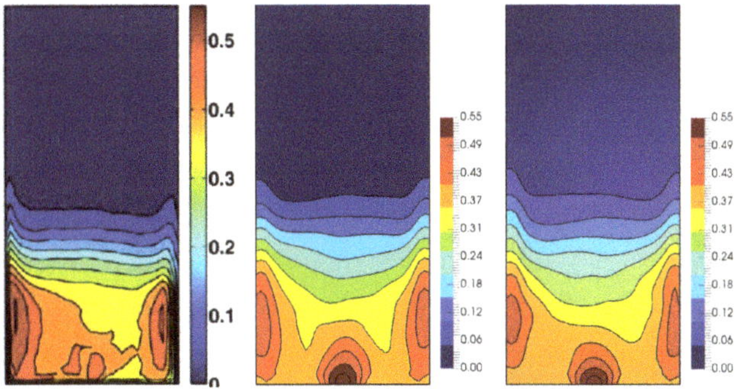

Figure A1. Time-averaged particle volume fraction. (**Left**): Experimental results, (**middle**): numerical results using Smagorinsky model, (**right**): numerical results without Smagorinsky model.

References

1. Tenneti, S.; Subramaniam, S. Particle-resolved direct numerical simulation for gas-solid flow model development. *Annu. Rev. Fluid Mech.* **2014**, *46*, 199–230. [CrossRef]
2. Pepiot, P.; Desjardins, O. Numerical analysis of the dynamics of two-and three-dimensional fluidized bed reactors using an Euler–Lagrange approach. *Powder Technol.* **2012**, *220*, 104–121. [CrossRef]
3. Alder, B.J.; Wainwright, T. Phase transition for a hard sphere system. *J. Chem. Phys.* **1957**, *27*, 1208–1209. [CrossRef]

4. Cleary, P.W. Industrial particle flow modelling using discrete element method. *Eng. Comput.* **2009**, *26*, 698–743. [CrossRef]
5. Deen, N.G.; Annaland, M.V.S.; Van der Hoef, M.A.; Kuipers, J.A.M. Review of discrete particle modeling of fluidized beds. *Chem. Eng. Sci.* **2007**, *62*, 28–44. [CrossRef]
6. Sundaresan, S. Reflections on mathematical models and simulation of gas-particle flows. In *Proceedings of the 10th International Conference on Circulating Fluidized Beds and Fluidization Technology*; Knowlton, T., PSRI, Eds.; ECI Symposium Series; Princeton University: Princeton, NJ, USA, 2011.
7. Jackson, R. *The Dynamics of Fluidized Particles*; Cambridge University Press: Cambridge, UK, 2000.
8. Capecelatro, J.; Desjardins, O. An Euler–Lagrange strategy for simulating particle-laden flows. *J. Comput. Phys.* **2013**, *238*, 1–31. [CrossRef]
9. Moureau, V.; Domingo, P.; Vervisch, L. Design of a massively parallel CFD code for complex geometries. *C. R. Mécanique* **2011**, *339*, 141–148. [CrossRef]
10. Chorin, A.J. Numerical solution of the Navier-Stokes equations. *Math. Comput.* **1968**, *22*, 745–762. [CrossRef]
11. Pierce, C.D.; Moin, P. Progress-variable approach for large-eddy simulation of non-premixed turbulent combustion. *J. Fluid Mech.* **2004**, *504*, 73–97. [CrossRef]
12. Moureau, V.; Domingo, P.; Vervisch, L. From large-eddy simulation to direct numerical simulation of a lean premixed swirl flame: Filtered laminar flame-pdf modeling. *Combust. Flame* **2011**, *158*, 1340–1357. [CrossRef]
13. Smagorinsky, J. General circulation experiments with the primitive equations: I. The basic experiment. *Mon. Weather Rev.* **1963**, *91*, 99–164. [CrossRef]
14. Germano, M.; Piomelli, U.; Moin, P.; Cabot, W.H. A dynamic subgrid-scale eddy viscosity model. *Phys. Fluids A Fluid Dyn.* **1991**, *3*, 1760–1765. [CrossRef]
15. Lilly, D.K. A proposed modification of the Germano subgrid-scale closure method. *Phys. Fluids A Fluid Dyn.* **1992**, *4*, 633–635. [CrossRef]
16. Cundall, P.A.; Strack, O.D. A discrete numerical model for granular assemblies. *Géotechnique* **1979**, *29*, 47–65. [CrossRef]
17. Ergun, S. Fluid Flow through Packed Columns. *J. Chem. Eng. Prog.* **1952**, *48*, 89–94.
18. Wen, C.Y.; Yu, Y.H. A generalized method for predicting the minimum fluidization velocity. *AIChE J.* **1966**, *12*, 610–612. [CrossRef]
19. Huilin, L.; Gidaspow, D. Hydrodynamics of binary fluidization in a riser: CFD simulation using two granular temperatures. *Chem. Eng. Sci.* **2003**, *58*, 3777–3792. [CrossRef]
20. Sun, R.; Xiao, H. Diffusion-based coarse graining in hybrid continuum–discrete solvers: Applications in CFD-DEM. *Int. J. Multiph. Flow* **2015**, *72*, 233–247. [CrossRef]
21. Mendez, S.; Gibaud, E.; Nicoud, F. An unstructured solver for simulations of deformable particles in flows at arbitrary Reynolds numbers. *J. Comput. Phys.* **2014**, *256*, 465–483. [CrossRef]
22. Kraushaar, M. Application of the Compressible and Low-Mach Number Approaches to Large-Eddy Simulation of Turbulent Flows in Aero-Engines. Ph.D. Thesis, Institut National Polytechnique de Toulouse, Toulouse, France, 2011.
23. Van der Vorst, H.A. Parallel iterative solution methods for linear systems arising from discretized pde's. In *Special Course on Parallel Computing in CFD*; France Workshop Lecture Notes; 1995. Available online: https://www.semanticscholar.org/paper/Parallel-Iterative-Solution-Methods-for-Linear-from-Vorst/907dfb316057751b85e8620cf02d4fecb9ed731d (accessed on 12 February 2019).
24. Van der Hoef, M.A.; Ye, M.; van Sint Annaland, M.; Andrews, A.T.; Sundaresan, S.; Kuipers, J.A.M. Multiscale modeling of gas-fluidized beds. *Adv. Chem. Eng.* **2006**, *31*, 65–149.
25. Lubachevsky, B.D. How to simulate billiards and similar systems. *J. Comput. Phys.* **1991**, *94*, 255–283. [CrossRef]
26. Fede, P.; Moula, G.; Ingram, A.; Dumas, T.; Simonin, O. 3D Numerical simulation and PEPT experimental investigation of pressurized gas-solid fluidized bed hydrodynamic. Presented at the ASME 2009 Fluids Engineering Division Summer Meeting, Vail, CO, USA, 2–6 August 2009.
27. Patil, A.V.; Peters, E.A.J.F.; Kuipers, J.A.M. Comparison of CFD-DEM heat transfer simulations with infrared/visual measurements. *Chem. Eng. J.* **2015**, *277*, 388–401. [CrossRef]
28. Lorenz, A.; Tuozzolo, C.; Louge, M.Y. Measurements of impact properties of small, nearly spherical particles. *Exp. Mech.* **1997**, *37*, 292–298. [CrossRef]

29. Goldschmidt, M.J.V.; Beetstra, R.; Kuipers, J.A.M. Hydrodynamic modelling of dense gas-fluidised beds: Comparison and validation of 3D discrete particle and continuum models. *Powder Technol.* **2004**, *142*, 23–47. [CrossRef]
30. Gorham, D.A.; Kharaz, A.H. Results of particle impact tests. In *Impact Research Group Report IRG 13*; The Open University: Milton Keynes, UK, 1999.
31. Yang, L.; Padding, J.J.; Kuipers, J.H. Investigation of collisional parameters for rough spheres in fluidized beds. *Powder Technol.* **2017**, *316*, 256–264. [CrossRef]
32. Hoomans, B.P.B.; Kuipers, J.A.M.; Briels, W.J.; van Swaaij, W.P.M. Discrete particle simulation of bubble and slug formation in a two-dimensional gas-fluidised bed: A hard-sphere approach. *Chem. Eng. Sci.* **1996**, *51*, 99–118. [CrossRef]
33. Li, T.; Gopalakrishnan, P.; Garg, R.; Shahnam, M. CFD–DEM study of effect of bed thickness for bubbling fluidized beds. *Particuology* **2012**, *10*, 532–541. [CrossRef]
34. Ishibashi, I.; Perry, C., III; Agarwal, T.K. Experimental determination of contact friction for spherical glass particles. *Soils Found.* **1994**, *34*, 79–84. [CrossRef]
35. Sandeep, C.S.; Senetakis, K. Effect of Young's modulus and surface roughness on the inter-particle friction of granular materials. *Materials* **2018**, *11*, 217. [CrossRef]

fluids

MDPI

Article

Modeling the Excess Velocity of Low-Viscous Taylor Droplets in Square Microchannels

Thorben Helmers [1,*], Philip Kemper [2], Jorg Thöming [2] and Ulrich Mießner [1]

[1] Department of Environmental Process Engineering, University Bremen, Leobener Str. 6, 28359 Bremen, Germany
[2] Department of Chemical Process Engineering, University Bremen, Leobener Str. 6, 28359 Bremen, Germany
* Correspondence: helmers@uvt.uni-bremen.de; Tel.: +49-421-218-63337

Received: 18 July 2019; Accepted: 26 August 2019; Published: 2 September 2019

Abstract: Microscopic multiphase flows have gained broad interest due to their capability to transfer processes into new operational windows and achieving significant process intensification. However, the hydrodynamic behavior of Taylor droplets is not yet entirely understood. In this work, we introduce a model to determine the excess velocity of Taylor droplets in square microchannels. This velocity difference between the droplet and the total superficial velocity of the flow has a direct influence on the droplet residence time and is linked to the pressure drop. Since the droplet does not occupy the entire channel cross-section, it enables the continuous phase to bypass the droplet through the corners. A consideration of the continuity equation generally relates the excess velocity to the mean flow velocity. We base the quantification of the bypass flow on a correlation for the droplet cap deformation from its static shape. The cap deformation reveals the forces of the flowing liquids exerted onto the interface and allows estimating the local driving pressure gradient for the bypass flow. The characterizing parameters are identified as the bypass length, the wall film thickness, the viscosity ratio between both phases and the *Ca* number. The proposed model is adapted with a stochastic, metaheuristic optimization approach based on genetic algorithms. In addition, our model was successfully verified with high-speed camera measurements and published empirical data.

Keywords: Taylor flow; droplet excess velocity; droplet velocity model; microfluidics; genetic algorithms; greybox modeling

---

1. Introduction

Microscopic multiphase flows facilitate a wide field of possible applications since they provide short diffusion layers within the flow structures. This enables high mass and heat transfer rates [1,2] for several applications ranging from extraction [3] and multiphase catalyst reactions [4] to improved unit operations such as mixing tasks [5]. The distinct features allow performing reactions at new process windows with fewer hazards or higher selectivity [6]. The specific flow conditions can furthermore serve for cell isolation [7], genetic analysis [8] and reaction screening in a droplet chain [9]. In contrast to large scale multiphase flows, microscopic flows are much easier to predict as there are no complex interactions such as swarm turbulence [10,11] commonly found in bubble columns. In fact, the reproducibility of Taylor flows is a key for the application of microscopic multiphase flows [12].

In the Taylor flow regime, the disperse phase is separated from the wall by a thin wall film and does not fill the entire cross-section of the microchannel. The remaining space between the droplet and the microchannel's corners is occupied by the continuous phase, which is referred to as gutters [13]. The droplets are typically longer than the channel diameter, which leads to separated elongated disperse phase instances. The continuous phase segments between the droplets are called slugs. Taylor flows are mainly established in circular capillaries or rectangular microchannels whereas especially the hydrodynamics of moving Taylor droplets in circular capillaries and the role of the thin wall film have been intensively studied [14].

Chemical reactions on the microscale are often performed in monolith reactors functioning as as a catalyst support. In these reactors, a high number of parallelized channels with hexagonal or square channel cross-section offer a high specific reaction area for wall placed catalysts at small wall thickness. This results in a better heat transfer through the walls and better mechanical stability than circular capillaries [15]. For process control and stabilization, as well as precise reactor design, knowledge of the underlying fluidic terms is crucial. The high grade of parallelity complicates the prediction of the hydrodynamics and the resulting pressure drop [16,17].

Besides disperse phase size distribution and formation frequency [18], the actual droplet velocity is essential for the droplet residence time in the reactor. It determines the contact time of the educts and influences the pressure drop of the reactor [19]. In parallel reactors, the exact knowledge of the pressure drop is especially necessary since a steady educt supply for each individual single reactor is needed to ensure stable and efficient working conditions [20].

Several publications deal with the droplet velocity in rectangular capillaries and observe a droplet velocity mostly faster than the superficial velocity (see Section 3). For flows in circular capillaries, where only a thin wall film is present, this velocity difference is well understood [21], while for rectangular microchannels a variety of explanations exist, which mostly correlate the relations from measurements [22]. This complicates the transfer of results to other flow applications or altered process parameters since local and instantaneous hydrodynamic parameters are mostly not taken into account by the models and correlations.

This work aims to establish a model to determine the droplet velocity from the actual flow conditions [23]: e.g., droplet length, material properties, and the Ca number. In a first step, we develop a concept for the relative droplet velocity, which bases this velocity on extrinsic parameters, allowing a linear scaling. From this concept, we identify the bypass flow through the gutters as well as the film-thickness as the prominent parameters for the excess velocity.

In the next step, we develop a model that uses the local surface curvature of the gutters to retrieve the local pressure at the entrance and outlet of the gutter as a driving force. The bypassing gutter flow is calculated based on the counterplay of this driving force, the gutter length and a viscosity correlated resistance factor β. The local droplet curvature at the gutter entrances is derived with an analytical interface shape approximation [24] and a correlation for the droplet cap curvature based on the Ca number from our previous work [25]. The model was successfully validated using high-speed camera measurements.

2. Hydrodynamic Fundamentals of Taylor Flows

The Taylor flow regime in rectangular microchannels is mainly influenced by surface tension forces rather than inertia forces. In the Taylor flow regime, a droplet fills nearly the whole cross-section of a hydrodynamic channel, while the continuous phase occupies the gutters and a thin wall film. The droplets are divided by the slugs, consisting only of liquid from the continuous phase.

The interaction between interfacial and viscous forces is described by the Capillary number Ca.

$$Ca = \frac{u_0 \eta_c}{\sigma} \tag{1}$$

Herein, η_c represents the dynamic viscosity of the continuous phase, σ the interfacial tension between both phases and u_0 the total superficial flow velocity:

$$u_0 = \frac{Q_0}{A_{ch}} = \frac{Q_d + Q_c}{W \cdot H} = \frac{Q_d + Q_c}{H^2 \cdot ar} \tag{2}$$

The superficial velocity u_0 is based on the volume flow of the disperse Q_d and continuous Q_c phase as well as the microchannel's cross-sectional area A_{ch} that calculates from the channel width W and channel height H or, respectively, the channel aspect ratio $ar = \frac{W}{H}$.

For energetic considerations, knowledge of the ratio between inertia and viscous forces is of importance. The Reynolds number

$$Re = \frac{u_0 H \rho_c}{\eta_c} \tag{3}$$

is used to describe the ratio between inertia and viscous forces of the continuous phase with ρ_c being the continuous phase density and η_c the continuous phase viscosity. Additionally, the viscosity ratio

$$\lambda = \frac{\eta_d}{\eta_c} \tag{4}$$

between both phases has a significant influence on the pressure drop and the velocity of a droplet [19,26,27] since it indicates the momentum coupling into the disperse phase. Please note that the definition of λ differs in the mentioned publications.

Considering the overall classification of the applied flow system, the material properties of both fluid phases are of importance. The Ohnesorge number Oh describes the most prominent material properties for droplets being formed or dispersed [28]. It characterizes the fluidic system independent of current flow or forces and is mostly used when working with surfactants to manipulate the flow properties. Here, we use the Oh number to characterize the continuous phase.

$$Oh = \frac{\eta_c}{\sqrt{d \rho_c \sigma}} = \sqrt{\frac{Ca}{Re}} \tag{5}$$

In many applications, Taylor droplets in rectangular microchannels move with a velocity different from the superficial velocity, because the droplet does not fill the entire channel cross-section and continuous phase can bypass the droplet through the gutters. An early description was given by Wong et al. [29], who described these phenomena analytically and declared two possible regimes.

In the first regime, the fluid in the gutters moves slower than the droplet, dissipating kinetic energy. For the gutters, Abiev [21] reported an inverted pressure gradient, indicated by the local surface curvature. For circular microchannels, this results in an inverse flow of the wall film and the droplet moves faster than the superficial velocity.

In the second flow regime, which holds true for long and highly viscous droplets [27,29,30], more energy is dissipated through the larger wall film area and through viscous dissipation in the droplets. Consequently, the flow in the gutters moves from the droplet back to the front. Thus, the droplet moves slower than the superficial velocity.

For both regimes, the thin wall film resists the motion of the droplet, resulting in a difference in pressure with higher value at the back and a lower value at the front of the droplet. The transition between both regimes is described by a critical flow rate and depends on the droplet length and the channel aspect ratio [29].

Jakiela et al. [31] focused experimentally on the influence of the momentum coupling between both phases represented by λ and also revealed a dependence of the droplet velocity on the droplet length. For short, low viscous droplets ($\lambda < 1$), the droplets move faster than the superficial velocity and the droplet behavior is assigned to the first flow regime. Highly viscous droplets ($\lambda > 1$) move either faster or slower than the superficial velocity, depending strongly on the droplet length.

3. Concept of Excess Velocity

Based on the instantaneous droplet velocity u_d, which is evident and directly measurable via optical or electrical measurement techniques [32], different explanatory approaches for the deviation of superficial velocity u_0 and droplet velocity u_d have been reported.

Liu et al. [33] defined this relative difference as slipping velocity u_{slip}, Howard and Walsh [34] as relative drift velocity u_{drift}, Angeli and Gavriilidis [35] as relative bubble velocity u_{rel} and Abadie et al. [36] as dimensionless droplet velocity. Jakiela et al. [31] focused directly on the ratio of $\frac{u_d}{u_0}$ and named this quotient droplet mobility according to Bretherton [37]. For this work, we summarize these approaches as a slipping velocity:

$$u_{slip} = \frac{u_d - u_0}{u_d} = 1 - \frac{u_0}{u_d} \tag{6}$$

In those concepts, the desired quantity is scaled with an intrinsic value such as the instantaneous droplet velocity, which leads to normalization effects as values $u_{slip} < 1$ and $u_{slip} > 1$ are not normalized symmetrically (Figure 1). This behavior has to be taken into account when experimental or simulative data is interpreted.

In our approach, we scale the velocity difference with the superficial velocity as an extrinsic property and, staying in the terms of extrinsic denomination, define it as an excess velocity u_{ex}

$$u_{ex} = \frac{u_d - u_0}{u_0} = \frac{u_d}{u_0} - 1 \tag{7}$$

In this manner, u_{ex} results values around 0 for droplet velocities equal to the superficial velocity (plug flow), while positive and negative values indicate droplets that are, compared to the superficial velocity, faster or slower, respectively.

The advantage of this extrinsic concept stands out in a comparison of both approaches (Figure 1). The first shown intrinsic concept (slip velocity) leads to an asymmetrical scaling behavior, especially for droplets with $u_d < u_0$, since $u_{slip} < 0$ decreases more strongly than $u_{slip} > 0$ would increase. For applications such as balancing or process modeling, a linear behavior of ratio values is preferable to simplify the balances. Otherwise, the nonlinear behavior would lead to an additional bias towards any direction $u_d < u_0 \vee u_d > u_0$, which requires a nonlinearly compensation in the description of the phenomena and complicates the model development.

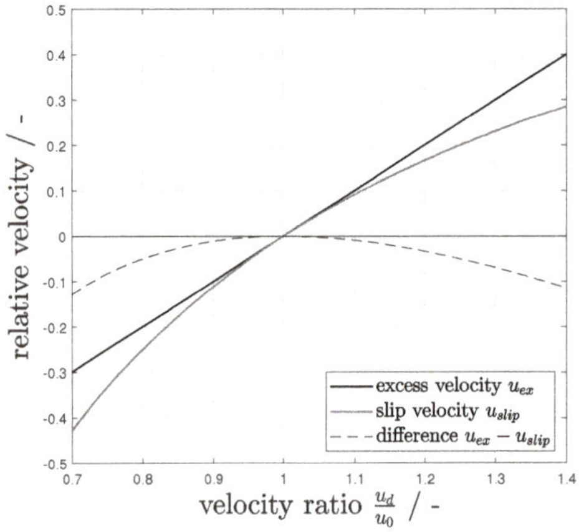

Figure 1. Comparison of different concepts of calculating the relative droplet velocity: excess velocity from this work (black solid line), slip velocity (grey solid line) and difference of both concepts (black dashed line).

A description for the excess velocity can be derived from the volume flows around moving droplets (Equation (8)). The continuity equation describes the connection between the gutter flow and the outer driving flows and delivers the relation between the total flow Q_0 and the volume flow fractions of the disperse Q_d as well as the continuous phase Q_c. Considering the unit cell of a single slug and an adjoining droplet, the continuous phase divides into the volume flow of the slug (Q_s), the flow through the gutters (Q_g), and through the wall film (Q_f):

$$Q_0 = Q_d + Q_c = Q_d + \left(Q_s + Q_g + Q_f\right) \tag{8}$$

If we depict the flows at a moving Taylor droplet (Figure 2a) and introduce a stationary control surface Γ (Figure 2b), velocities can be retrieved from the balances, while two different flow states are possible: In the case of a droplet passing Γ, the slug volume flow through the control surface is $Q_s = 0$ (since there is no slug present). For a slug passing Γ, only the slug volume Q_s is present. As one can see, the use of a stationary point of view leads to nonstationary terms in the balances.

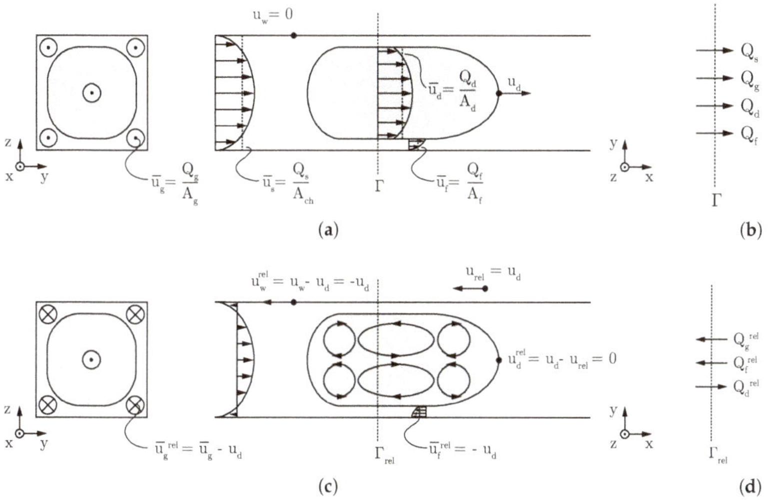

Figure 2. Prominent averaged and local velocity for a flowing droplet, where overlined entities represent area-averaged velocities. The film flow is not shown in this drawing. (**a**) Velocities for a fixed point-of-view (fixed frame); (**b**) flow balance at a steady control surface; (**c**) velocities for a moving point-of-view with the velocity u_d (moving frame); and (**d**) flow balance at a moving control surface.

If the coordinate system is changed from a fixed frame to a moving frame by moving the control surface with an arbitrary velocity u_{rel}, the balances become stationary and relative velocities become visible (Figure 2c). For this moving coordinate system, an additional volume flow Q_{rel} adds to the balances that result from the transformation of the coordinate system (Figure 2d):

$$Q_{rel} = u_{rel} \cdot A_{ch} \tag{9}$$

This changes Equation (8) to:

$$Q_0 - Q_{rel} = Q_d + Q_g + Q_f - Q_{rel} \tag{10}$$

With knowledge of the specific areas for each distinct volume flow rate, the averaged velocities can be calculated. Following Figure 3a,c, we conclude for the gutter area A_g of all four gutters

$$A_g = 4\left[\left(\overline{R_g}^2 - \frac{\pi \overline{R_g}^2}{4}\right) + 2\delta\overline{R_g} + \delta^2\right] \tag{11}$$

with δ denoting the wall film-thickness and $\overline{R_g}$ representing the mean dynamic gutter radius, which is described below in Section 4. For the cross-sectional film area A_f, we state with the aspect ratio $ar = WH^{-1}$

$$A_f = \frac{4\delta}{H}H^2\left(\frac{1+ar}{2} - 2\frac{\overline{R_g}+\delta}{H}\right) \tag{12}$$

The droplet cross-sectional area A_d is delivered combining A_g and A_f:

$$A_d = A_{ch} - \left(A_g + A_f\right) \tag{13}$$

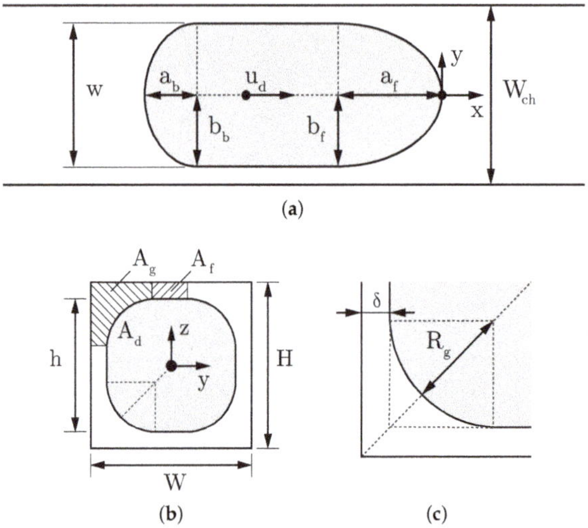

Figure 3. Declaration of relevant geometry for the model of a Taylor droplet flowing through a rectangular microchannel with the droplet velocity u_d. (**a**) Top-view of x-y-plane, characterizing the droplet with the front (a_f, b_f) and back cap (a_bf, b_b), as well as the channel width W and droplet width w. (**b**) Droplet front-view in y-z-plane with the droplet area A_d, gutter area A_g, film area A_f, channel height H and the droplet height h. Only one representation of each area is shown. (**c**) Close-up of the droplet corner region with gutter radius (R_g) and the film-thickness δ.

With the cross-sectional areas of the droplet, the gutter and the wall film, the volume flow rates can be rearranged to area-averaged velocities and Equation (10) becomes

$$u_0 A_{ch} - u_{rel} A_{ch} = \overline{u_d} A_d + \overline{u_g} A_g + \overline{u_f} A_f - u_{rel} A_g - u_{rel} A_d - u_{rel} A_f \tag{14}$$

Herein, $\overline{u_d}$, $\overline{u_g}$ and $\overline{u_f}$ are area-averaged velocities of the droplet, gutter and film, respectively.

For the droplet ($Q_d = \overline{u_d} A_d - u_{rel} A_d$) and the film volume flow ($Q_f = \overline{u_f} A_f - u_{rel} A_f$), the transition velocity u_{rel} equals the stagnation point velocity u_d, since we assume incompressibility, mass conservation, a stagnant film and a stationary droplet shape [38].

$$\overline{u_d} = u_d \stackrel{!}{=} u_{rel} \tag{15}$$

Additionally, we assume the averaged velocity in the thin wall films to be insignificant for $Ca < 0.2$

$$\overline{u_f} \approx 0 \tag{16}$$

Therefore, Equation (14) simplifies to

$$(u_0 - u_d)A_{ch} = \overline{u_g^{rel}}A_g + (0 - u_d)A_f \tag{17}$$

Herein, $\overline{u_g^{rel}} = \overline{u_g} - u_d$ represents the relative gutter velocity, which dissipates flow energy in the gutter. Simplification and combination with Equation (7) leads to

$$\frac{u_d}{u_0} - 1 = u_{ex} = -\frac{\overline{u_g^{rel}}A_g}{u_0 A_{ch}} + (u_{ex} + 1)\frac{A_f}{A_{ch}} \tag{18}$$

We neglect the terms that are small of higher-order (see Appendix A) and retrieve an expression for the excess velocity:

$$u_{ex} = -\frac{Q_g^{rel}}{Q_0} + \frac{A_f}{A_{ch}} \tag{19}$$

The relative gutter volume flow (Q_g^{rel}) and the cross-sectional area of the wall film A_f are the most prominent influencing quantities for the excess velocity. Thus, our proposed model aims to especially determine these quantities.

4. Model Specification

The considerations of the previous section identify the volume flows, which determine the excess velocity. In a second step, we clarify the relevant influential parameters on these volume flows and their interconnection.

For the proposed modeling approach, we adapt a greybox model following Hangos and Cameron [39]. Our model is developed from engineering principles, hydrodynamic considerations (see Section 3) and well-defined equations, whereas the initialization of a part of the influential parameters is based on measured data. The underlying relations can be described as an intermediate concept between a black box (completely based on measurement data) and a white box model (based only on analytically well-known equations and engineering principles).

As depicted in Section 3, we assume a droplet flowing through a rectangular microchannel with its properties according to Figure 3. The thin wall film cross-section A_f is determined by Equation (12) and depends on the channel height H, the channel aspect ratio ar and the film-thickness δ. To determine δ, we apply the model of Han and Shikazono [40], which holds for $Ca < 0.2$. For the proposed excess velocity model, full wettability of the walls with a stable wall film is assumed. Besides this behavior being proven in the later measurements (see Section 6), we compared this assumption with the literature. Kreutzer et al. [41] analytically retrieved the counterplay of film rupture in the region near the gutters and film formation at the bubble or droplet front. Furthermore, they found coherences for the length where the film starts to rupture. For short or fast droplets, new films are formed more quickly than they rupture. In this work, mainly short droplets are covered.

Khodaparast et al. [42] focused on the dewetting of the wall film and found a distance from the droplet front, at which dewetting occurs, to correlate with Ca in the range $10^{-5} \leqslant Ca \leqslant 10^{-2}$ and the surface wettability. As a criterion, they found a minimum film-thickness of 100 nm, beyond which rupture is suppressed. For the presented range of the model ($10^{-5} \leqslant Ca \leqslant 10^{-2}$, $0.5 \leqslant Re \leqslant 3$), based on the film thickness model of Han and Shikazono [40], the critical film-thickness never drops below this value.

For the flow of the bulk phase as well as the flow in the gutters, we omit the influence of viscous heating based on the criterion of Morini [43]. We elaborate on this in Appendix B.

The relative volume flow through the gutter Q_g^{rel} is derived from the pressure difference along the gutters, as suggested by Abiev [21]. Therefore, knowledge of the relevant pressures is crucial. It can be determined from the dynamic interface deformation caused by the moving liquids through the gutters in flow direction [44]. Stagnant droplets have a static cap shape with a circular outline according to

Musterd et al. [45]. When set into motion, the moving liquids exert forces onto the interface and cause a dynamic shape deformation [38].

A model proposed by Mießner et al. [24] allows to approximate the droplet shape and gutter diameter. The model implies that the deformation difference of the dynamic droplet cap shape between the droplet front and back results in a change of the gutter radius from the static shape. The cross-sectional gutter area A_g widens asymmetrically from back to front. The growing gutter radii accommodate the relative volume flow Q_g^{rel} of the continuous phase through the gutter. The gutter entrance at the droplet front is therefore larger than the gutter exit at the droplet back. Utilizing their model, the dimensionless radius of these gutters can be calculated from the flow-related curvature of the droplet.

The gutter radius $k_{g,i}$ is therefore defined as a fraction of the droplet height h. For the case of present wall films, the term is expressed as follows:

$$k_{g,i} = \frac{R_{g,i}}{h} = \frac{R_{g,i}}{H - 2\delta} \tag{20}$$

To simplify the geometry, we define a mean gutter radius $\overline{R_g}$ using Equation (20):

$$\overline{R_g} = \frac{k_{g,f}(H - 2\delta) + k_{g,b}(H - 2\delta)}{2} \tag{21}$$

$$= \frac{(H - 2\delta)(k_{g,f} + k_{g,b})}{2} \tag{22}$$

which is also used for the mean cross-sectional area A_g derived with Equation (11).

In previous work [25], we introduced a quantification of the droplet cap deformation with an elliptic approximation of the cap outline

$$k_{c,f} = \frac{a_f}{b_f} \tag{23}$$

$$k_{c,b} = \frac{a_b}{b_b} \tag{24}$$

The ratio of the semi-major a and semi-minor axis b of the droplet cap curvature is introduced as deformation ratios $k_{c,i}$ at the droplet front and back. They become $k_{c,f} = k_{c,b} = 1$ when describing the static circular droplet cap shape of the droplet front and back. For the dynamic cap shape, under the influence of the moving liquids, the droplet front appears elongated $k_{c,f} > 1$ and the back cap is compressed in flow direction $k_{c,f} < 1$. Hence, we introduce a correlation to describe these relations: The cap curvature only depends on the Ca number for moderate flows ($Re < 5$):

$$k_{c,i} = m_{cap} \cdot Ca^{c_{cap}} + n_{cap} \tag{25}$$

With the correlation approach from our recent publication and the model from Mießner et al. [24], we are able to calculate the Laplace pressure at the gutters. We assume a linear connection between the gutter front and the back of the droplet body, since the curvature of the gutter in flow direction is negligibly small. In this case, the mean interface curvature at the gutter entrance and its exit depends on the gutter radii only, and the flow-induced Laplace pressure difference equals:

$$\Delta p_{g,f} = \sigma \left(\frac{1}{R_{g,f}^{stat}} - \frac{1}{R_{g,f}} \right) \tag{26}$$

$$\Delta p_{g,b} = \sigma \left(\frac{1}{R_{g,b}^{stat}} - \frac{1}{R_{g,b}} \right) \tag{27}$$

Those deformation related pressures at the droplet front and back provide a link to the driving pressure difference Δp_{LP} along the gutter length in the flow direction. Due to symmetry, the static terms cancel out:

$$\begin{aligned} \Delta p_{LP,fb} &= \Delta p_{g,f} - \Delta p_{g,b} \\ &= \sigma \left(\frac{1}{R_{g,f}} - \frac{1}{R_{g,b}} \right) \end{aligned} \tag{28}$$

Using the dimensionless expression from Equation (20), this results in

$$\Delta p_{LP,fb} = \sigma \left(\frac{1}{k_{g,f}\,(H - 2\delta)} - \frac{1}{k_{g,b}(H - 2\delta)} \right) \tag{29}$$

$$= \left(\frac{\sigma}{H} \right) \frac{1}{\left(1 - \frac{2\delta}{H}\right)} \left(\frac{k_{g,b} - k_{g,f}}{k_{g,f} \cdot k_{g,b}} \right) \tag{30}$$

for the flow-induced pressure difference as a driving force.

The relative volume flow rate through the four gutters Q_g^{rel} can be modeled as a laminar pressure driven flow [46] and it is linked to this pressure difference with a hydrodynamic resistance Ω:

$$Q_g^{rel} = \frac{1}{\Omega} \Delta p_{LP,fb} \tag{31}$$

The hydrodynamic resistance Ω was defined by Ransohoff and Radke [47] and Shams et al. [27] as

$$\frac{1}{\Omega} = \frac{\overline{R_g}^2}{\beta} \frac{A_g}{\eta_c \overline{l_g}} \tag{32}$$

Besides the gutter length l_g, β denotes a dimensionless factor, which represents the geometrical obstructions of the gutter flow, as well as the viscous coupling of both flow phases. In accordance with the simulation results of Shams et al. [27], we declare an influence from the viscosity ratio λ of the flow phases to take care of the viscous coupling effects:

$$\beta = m_\beta \cdot \lambda^{c_\beta} + n_\beta \tag{33}$$

The gutter length l_g can be derived from the droplet length l_d, if the gutter distance $\Delta x_{g,i}$ from the caps is subtracted:

$$l_g = l_d - \Delta x_{g,f} - \Delta x_{g,b} \tag{34}$$

The gutter distance from the front and back droplet tip $\Delta x_{g,i}$ was defined by Mießner et al. [24] as

$$\Delta x_{g,i} = k_{c,i} \frac{H}{2} \left[\left(ar - \frac{2\delta}{H} \right) + \left(1 - \frac{2\delta}{H} \right) (1 - 2k_{g,i}) \right] \tag{35}$$

The above-stated considerations lead to our final description for the relative volume flow through the gutters

$$Q_g^{rel} = \frac{\overline{R_g}^2}{\beta} \frac{A_g}{\eta_c \overline{l_g}} \Delta p_{LP,fb} \tag{36}$$

Herein, $\overline{l_g}$ and Δp_{LP} depend on $\overline{R_g}$ as the preceding considerations show. Thus, $\overline{R_g}$ has the most prominent influence feature besides β. Inserting Equations (36) and (33) into Equation (19) delivers

$$u_{ex} - \frac{A_f}{A_{ch}} = \frac{1}{Ca}\frac{1}{\beta}\frac{\overline{A_g}}{A_{ch}}\frac{\overline{R_g}^2}{l_g H}\frac{1}{(1-\frac{2\delta}{H})}\left(\frac{k_{g,b}-k_{g,f}}{k_{g,f}\cdot k_{g,b}}\right) \tag{37}$$

expanding with $\frac{W}{W}$ and inserting l_g (Equation (34)), we receive our final expression for the excess velocity:

$$u_{ex} = \frac{1}{Ca}\frac{1}{\beta}\frac{W}{l_d}\frac{\overline{A_g}}{A_{ch}}\frac{\overline{R_g}^2}{\left(1-\frac{\Delta x_{g,f}}{l_d}-\frac{\Delta x_{g,b}}{l_d}\right)A_{ch}}\frac{1}{(1-\frac{2\delta}{H})}\left(\frac{k_{g,b}-k_{g,f}}{k_{g,f}\cdot k_{g,b}}\right)+\frac{A_f}{A_{ch}} \tag{38}$$

We point out that our model is usable within the capillary regime $Ca < 0.02$ since the model for the wall film-thickness from Han and Shikazono [40], the analytic interface model from Mießner et al. [24] and the droplet curvature correlation from our recent work [25] are valid in this range. At higher Ca in the viscous regime, different flow conditions with thicker wall films [22] as well as a higher influence of Re are reported [48].

5. Model Calibration

The final expression for the excess velocity (Equation (38)) depends on accessible data such as W, ar and l_d as well as on $\overline{R_g}$, β, $k_{g,f}$, $k_{g,b}$, $\Delta x_{g,f}$ and $\Delta x_{g,b}$, whereas the latter is also related to the gutter radii $\overline{R_g}$. As shown in the previous section, the parameters can be calculated from the droplet cap curvatures $k_{c,f}$ and $k_{c,b}$ and via measurement-based model calibration. The six parameters ($m_{cap,f}$, $m_{cap,b}$, $n_{cap,f}$, $n_{cap,b}$, $c_{cap,f}$, and $c_{cap,b}$) influencing the cap deformation at the droplet front and back as well as the dimensionless resistance factor β with m_β, c_β and n_β need to be adjusted.

For model calibration, we used the dataset presented in [25] in combination with supplementary measurements in the data range of low Ca and redefine n_{cap} within an interval around 1. Nevertheless, for $Ca \to 0$, it represents the point of minimal surface energy and equates a spherical droplet shape. This approach allows improving the convergence of solvers. Out of further a priori considerations ($\beta > 0$, $n_{c,f} \approx 1$), we additionally defined the boundaries for the search space of the solver presented in Table 1 and allowed the solver to adapt the correlation coefficients from our last work to the presented model and acquired measurement data.

Table 1. Boundaries of input values for optimization.

Parameter	Lower Boundary	Upper Boundary
$m_{cap,f}$	1.00	9.90
$c_{cap,f}$	0.40	1.50
$n_{cap,f}$	1.00	1.005
$m_{cap,b}$	−2.50	−1.00
$c_{cap,b}$	0.30	0.75
$n_{cap,b}$	0.995	1.00
m_β	0.00	10.00
c_β	0.50	1.5
n_β	0.50	20

Besides the fixed boundaries of the search space, a hydrodynamic boundary condition was applied to improve the convergence of the used optimization algorithms. For a rising Ca, the difference between $k_{g,f}$ and $k_{g,b}$ must increase, because of the pressure difference between the droplet front and back increases with higher Ca [21] and the droplet front elongates, which leads to a larger front gutter radius, while the droplet rear flattens out. This is expressed by the gutter radius increase for an increasing Ca:

$$\frac{dk_{g,f}}{dCa} > \frac{dk_{g,b}}{dCa} \tag{39}$$

The large number of influence parameters leads to a highly nonlinear optimization problem with numerous local minima. Thus, most gradient-based algorithms are not suitable for this type of optimization problem, which tend to converge to local optima. This would result in an enormous number of randomly initialized solver calls to cover the whole search space. In contrast, stochastic and metaheuristic approaches cover the search space to solve the statistical part of the greybox model by finding the global optimum.

The quality of a solver result (e.g., deviation between measured data and estimation) is quantified by the loss-function of the problem. For the model calibration, we define

$$\mathcal{L} = \omega_1 \left(\sum \frac{|k_{c,b} - k_{c,b}^{\mathcal{M}}|}{k_{c,b}^{\mathcal{M}}} + \sum \frac{|k_{c,f} - k_{c,f}^{\mathcal{M}}|}{k_{c,f}^{\mathcal{M}}} \right) + \omega_2 \sum \frac{|u_d - u_d^{\mathcal{M}}|}{u_d^{\mathcal{M}}} + \omega_3 \sum \frac{|\psi - \psi^{\mathcal{M}}|}{\psi^{\mathcal{M}}} \tag{40}$$

with the weights of the individual properties $\omega_1, \omega_2, \omega_3$ following Table 2. Values with an upper index $\mathcal{M}$ denote values estimated by the model and values without an upper index represent the calibration data. The first two sums serve as calibration datasets for the hydrodynamic flow properties, since they contain the flow related deformation and therefore the hydrodynamic influences. The differences of the velocity u_d serves as a parameter for the actual droplet velocity and therefore the flow resistance β. This is necessary, since otherwise no representation for β is available.

Table 2. Weight factors ω_i of the loss-function for GA optimization.

ω_1	ω_2	ω_3
3.0	5.0	3.7

In addition, the concept of an equilibrium function ψ is introduced to maintain the overall integrity of the model: The excess velocity depends on extrinsic measurable values such as the dimensionless quantities and geometrical properties as well as varying flow properties such as the gutter length. Thus, it is appropriate to separate the measurable quantities from the model based quantities. In doing so, one can directly compare the quantities received from our correlation with measurement data adjusted for material and flow properties. The separation of those terms leads to the equilibrium function ψ (Equations (41) and (42)). Herein, Equation (41) represents the data from our measurements and Equation (42) only data from our modeling assumptions and geometry. For a well-adjusted model, the measured data for ψ (Equation (41)) should systematically correspond with the modeled values for ψ (Equation (42)).

$$\psi = \left(u_{ex} - \frac{A_f}{A_{ch}} \right) Ca\, \beta \frac{l_d}{W} \tag{41}$$

$$\psi^{\mathcal{M}} = \frac{\overline{A_g}}{A_{ch}} \frac{\overline{R_g}^2}{\left(1 - \frac{\Delta x_{g,f}}{l_d} - \frac{\Delta x_{g,b}}{l_d}\right) A_{ch}} \frac{1}{(1 - \frac{2\delta}{H})} \left(\frac{k_{g,b} - k_{g,f}}{k_{g,f} \cdot k_{g,b}}\right) \tag{42}$$

The calibration procedure is based on a Genetic Algorithm (GA). For good optimization results, it is necessary to emulate a sufficiently sized genome pool, thus a high number of emulated individuals is preferred. This in turn results in a massively increased calculation demand, because, for each iteration step, every single individual and their children need to be evaluated [49]. Additionally, the genetic algorithm is typically performed iteratively to identify local optima. To decrease the number of time-consuming iteration steps of the GA and thereby reduce overall calculation time, we utilize a three-step stochastic and gradient-free approach.

In the first step, a random population at the feasible borders of the problem is generated for the Genetic Algorithm and a genetic optimization performed. The convergence point of the GA is initialized via Latin Hypercube Sampling processed by a following fast-converging Pattern

Search Algorithm (PSA) [50] that results in an improved minimum as the final convergence point. The properties of both algorithms are shown in Table 3. The algorithm finally merges at the values shown in Table 4. The results are discussed in the following.

Table 3. Properties of used solver algorithms.

Genetic Algorithm	
population size	200 individuals
creation function	random feasible population
scaling function	ranking
selection function	stochastic uniform
mutation function	adaptive feasible
crossover function	scattered
Pattern Search Algorithm	
search method	Latin Hypercube
poll method	complete poll

Table 4. Values for the fit-functions for droplet shape k_i and resistance β.

Target Value	m_i	c_i	n_i
k_f	4.8761	0.7465	1.0021
k_b	−1.6967	0.4745	0.9980
β	6.1280	1.2105	1.4541

For the flow-induced cap curvature, the solver merges to an exponential behavior as a function of the *Ca* number for both cap deformation ratios k_f and k_b, which coincides with the results of our previous works [24,25,51]. The retrieved correlation with underlying data points is shown in the Appendix (Figure A2). Our additional boundary condition (Equation (39)) is satisfied for all values as the graphs for the gutter radii $k_{g,i}$ show.

For the resistance factor β, no fitting data are available since it cannot be measured directly. Therefore, β is fitted based on the velocity data from high-speed camera measurements and the application of parameters for the droplet deformation (measurements are performed in the experimental setup described in [25]). The resulting droplet velocities in comparison with the measured values are shown in Figure 4. The measurements fit reasonably well in the range of inevitable velocity fluctuations caused by Taylor flow stability of ±10 % range described by [13].

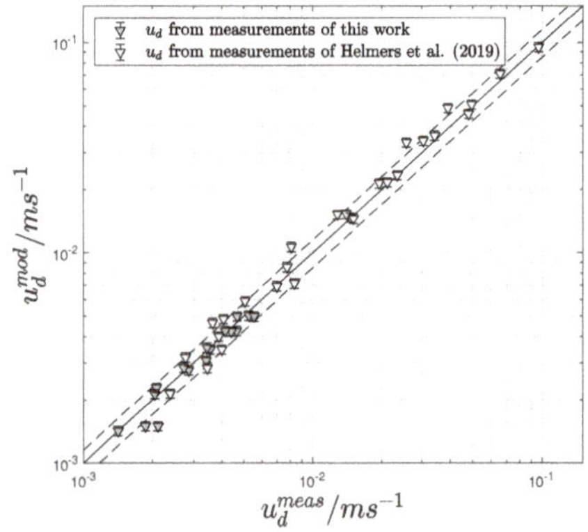

Figure 4. Parity plot for the droplet velocity of measured droplets and corresponding data from our model after adaption of the dimensionless resistance factor β. The values situate fairly good into the fluctuation range of 10%, as reported by van Steijn et al. [13], Fuerstman et al. [52].

6. Model Validation

For a first validation step, the equilibrium functions ψ and $\psi^{\mathcal{M}}$ are considered. In the case of a hydrodynamic well-adjusted model, both functions should coincide, and our model based shape deviations ($\psi^{\mathcal{M}}$) equal the combination of measured properties (ψ). The corresponding data and the values for our model agree well at high *Ca* numbers (Figure 5), whereas a deviation for lower *Ca* numbers can be observed. This can be explained by the fact that the excess velocity itself is a relative quantity and it is strongly influenced at lower absolute values (low *Ca* number). Thus, an inevitable constant measurement deviation for velocity and volume flows caused by the experimental equipment results in a higher error for low excess velocities. Especially for $Ca < 10^{-4}$, the resulting volume flows are situated at $Q_0 \approx 2 \times 10^{-6}$ L/min and even minor deviations lead to high errors for the excess velocity. Thus, we consider our model approach to represent the measurements reasonably well and our fit coefficients to be valid.

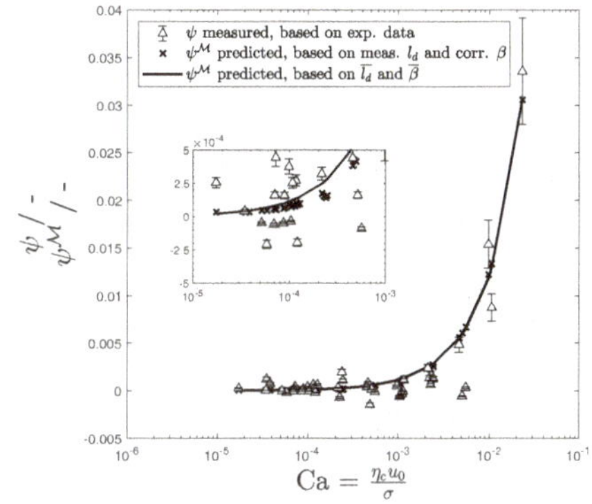

Figure 5. Calibration function ψ for measured (triangles) and modeled data $\psi^{\mathcal{M}}$ for the measured droplet lengths l_d and correlated β (crosses) over the *Ca* number. The black solid line depicts the calibration function for an averaged $\overline{l_d}$ and $\overline{\beta}$ to clarify the underlying systematic.

Besides the hydrodynamic validation, our assumptions for β are compared with available simulation data. We found a dependence of the gutter flow resistance β on the viscosity ratio λ. The latter can be interpreted as an indicator for the viscous coupling of both flow phases (Figure 6). In the case of a highly viscous continuous phase ($\lambda < 1$), the strongest velocity gradients are found inside the droplet, while, for a viscous disperse phase ($\lambda > 1$), larger velocity gradients and therefore energy dissipation are found inside the gutter-flow, resulting in a larger β.

This approach agrees with the simulations from Shams et al. [27], who improved the model of Ransohoff and Radke [47] by introducing the viscous coupling of the disperse and the continuous phase. For our case of $0.1 \leqslant \lambda \leqslant 1.4$ and a contact angle $\theta = 0°$, Shams et al. [27] reported a β of 20–30 for a co-current flow. Our values are shifted by a constant offset, while the slope and therefore the dependence on λ is similar. We consider the difference in the offset to be caused by the use of a different flow field specifications. Shams et al. [27] described a concurrent flow in a fixed frame specification, while in this work we determine Q_g^{rel} within a moving frame flow specification. The coordinate transformation only changes the offset of the function, while the hydrodynamic influence (the slope) must remain identical. Additionally, within the simulation of [27], they assumed a contact line between disperse phase, continuous phase and the wall in their problem definition. Although for the solution shown in Figure 6 the contact angle for the continuous phase is nearly 180°, the existence of a contact line introduces an additional resistance. Thus, regarding β, we consider our model plausible.

Due to the mentioned experimental restrictions, we cannot directly compare the modeled and experimental determined excess velocities u_{ex}. Instead, we compared the results of our model to different published approaches. A suitable correlation for the prediction of the excess velocity in the first regime was introduced by Jose and Cubaud [22], who identified the Ca number and the ratio of the droplet length $\frac{l_d}{H}$ as characteristic properties (Figure 7). It has to be mentioned that the model of Jose and Cubaud [22] for non-wetting droplets ends at $\frac{l_d}{H}Ca^{-1} > 600$, since for larger values they observed the disperse phase to wet the channel walls. This results in intensified dissipation and a higher pressure drop and thereby inhibits a gutter flow from the droplet front to the back. This phenomenologically equals the second flow regime, as mentioned by Wong et al. [5], but lacks the thin wall film and results in a much larger pressure drop. Furthermore, the viscosity ratio λ is not included in their correlation.

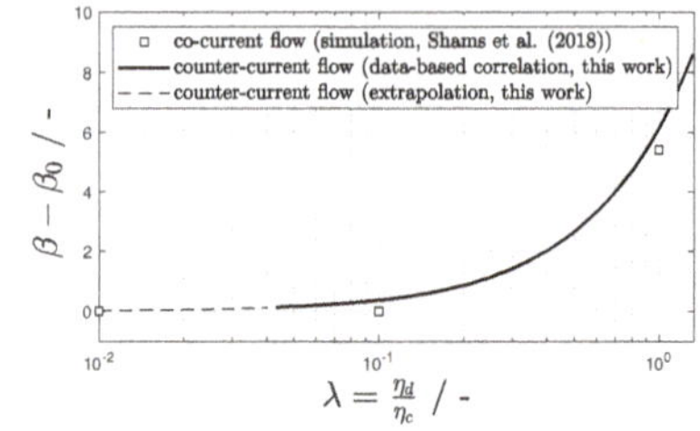

Figure 6. Comparison of calculated resistance factor β (squares) and correlation (solid and dashed line) of our model. The data of the simulation from Shams et al. [27] is shown as squares. Both datasets are corrected by the offset β_0 to compensate for the influence of the flow form (co-current or counter-current).

A comparison of our greybox model with the correlation from Jose and Cubaud [22] (Figure 7) shows good agreement. At very low Ca numbers ($Ca < 10^{-4}$), our model results in a slightly increased excess velocity. We regard this behavior of the model as not physical. The effect results from the mathematical counterplay of the terms $\lim_{Ca \to 0} \frac{1}{Ca} = \infty \leftrightarrow \lim_{Ca \to 0} k_{g,f} - k_{g,b} = 0$ within Equation (38). To achieve a stable solver convergence, we accepted a small residual deviation for the static case $Ca \to 0$ for the front and back shape of 0.1%. Due to the relative character of the excess velocity, this unfolds a significant influence at low Ca values. Unfortunately, additional measurement validation concerning the shape deviation at very low Ca is not possible in our experimental design, since the expected shape deviation is smaller than the blurriness of the interfacial area in the images itself and therefore lies inside the measurement deviation.

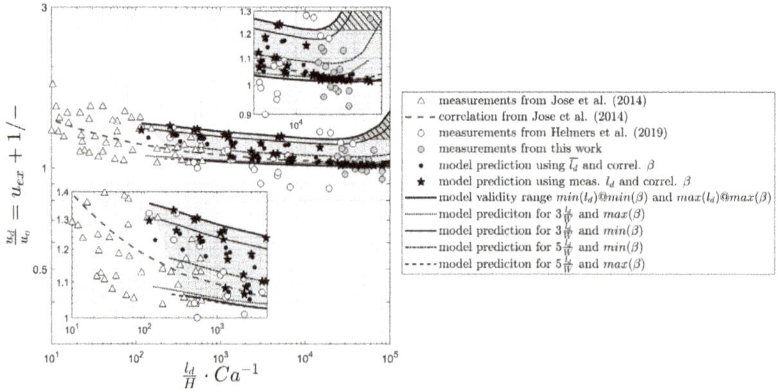

Figure 7. Comparison of our model (stars) and measurements (circles) with the measurements (triangles) and correlation from Jose and Cubaud [22] for a Taylor droplet in co-flow (axis scaling and normalization kept for comparability). The inclination for low Ca numbers (hatched area) is discussed in the text. Since the influence of l_d correlates linearly with Δp_{LP} instead of Ca in our model and β is not included in the x-axis normalization, we additionally show the borders of our model for the minimum/maximum l_d and β of our measurements.

7. Discussion

The interfacial area of a Taylor droplet in rectangular channels can be divided into the front and back cap regions, the wall films and the gutter interface. Neglecting the caps, the main momentum input into a droplet is transferred across the wall film and the gutter interface area. An increasing channel aspect ratio and droplet length result in a growing wall film area, i.e., an enlarged dissipation interface.

As shown in Section 2, the behavior of the Taylor droplet's excess velocity can be parted in two possible regimes and the viscosity ratio λ has a strong influence on the hydrodynamic mechanisms. In the first regime, the fluid in the gutter flows slower than the droplet, exerting a drag force and leading to a positive excess velocity $u_{ex} > 0$. These drag forces influence the droplet shape, leading to a flattened droplet back and elongated droplet front. We characterize this shape variation with a correlation (Section 5). As can be seen from the measurements, our data fall into this flow regime, as our resulting droplet shape indicates in accordance with Wong et al. [29].

The influence of the droplet length in the plug flow regime and $\lambda < 1$, where larger droplets have an $u_{ex} \approx 0$, is in agreement with Jakiela et al. [31] and their later publication [53]. Additionally, they found an elongation of the dynamic droplet length in comparison to the static droplet length, which is also covered by our shape correlation, since for rising the *Ca* number the droplet front elongates stronger than the droplet back is compressed. Recently published simulations by Kumari et al. [54] show that also for larger *Re* numbers $u_{ex} \approx 0$.

Our concept of deriving the excess velocity from the gutter pressure drop that is directed against the flow direction (larger pressure at the front gutter entrance) is in accordance with Abiev [21]. Nevertheless, the averaged pressure is still higher at the droplet back than at the droplet front due to the overall droplet pressure drop, since the droplet needs a driving force for its translation.

For the second flow regime identified by Wong et al. [29], where the viscous dissipation in the film and droplet leads to a bypass flow from the droplet back to the front and therefore a negative droplet excess velocity (for large λ and long droplets), our model can be adapted, if the gutter-shape-difference term is revised or a resistance coefficient for the film is added. As the viscosity ratio λ rises, more momentum will be dissipated via the wall films. This extra momentum is dissipated at the gutter interface, which results in a slower droplet velocity, forcing the continuous phase to bypass the droplet reversely. As the hydrophilized channels walls in our experimental setup did not allow establishing a water-in-oil two-phase flow with a $\lambda \gg 1$, droplet shape correlations, as well as measurements of u_{ex} for the case of $\lambda \gg 1$ should be performed in future work. Although we assume a systematic inversion of the gutter radii ratio back to front as a consequence of the reversed gutter flow direction, it should be mentioned that, with the current shape correlation, our model only works for the case of low viscous disperse phase ($\lambda \lessapprox 1$) such as gas/liquid or low viscous oil/water flows.

The influence of the channel aspect ratio, as mentioned by Wong et al. [5], is incorporated in our model: a higher aspect ratio results in lower excess velocities. This can be explained by the larger drag forces acting on a larger relative wall film area caused by the flattened channel geometry.

8. Conclusions

In this work, we developed a model to determine the relative droplet velocity of Taylor flows in square microchannels for moderate *Ca* numbers and low to moderate viscosity disperse phase ($\lambda \lessapprox 1$). We base our model on the relative volume flow through the gutters, as well as the wall film-thickness. The flow through the gutters is determined from the pressure drop described by the Laplace pressure difference between the gutter entrances.

Our model uses the gutter radii to obtain the resulting pressure gradient that drives the continuous phase through the gutters. We used measurements at different *Ca* and *Re* in a surfactant-free fluid system from a previous publication [25] to derive the radii at the gutter entrances from the surface shape model proposed by Mießner et al. [24] and calibrate the model parameters.

Our model was successfully validated with an intrinsic approach comparing the congruence of measurement data and calibrated model parameters. Additionally, we successfully compared our model to the phenomenological correlation of Jose and Cubaud [22].

The comparison with the most prominent approaches shows that our model and the chosen influential parameters are valid for moderate and small viscosity ratios. The excess velocity is determined by viscous dissipation in the droplet and the gutters as well as the drag of the thin wall films. The relation is characterized by the Ca number, the viscosity-ratio λ, the dimensionless gutter-length l_g, the aspect ratio ar and the wall film area A_f. Furthermore, the proposed model can close the gap for $\frac{l_d}{H}Ca^{-1} > 600$ and allows the calculation of the excess velocity for moderate Ca numbers ($Ca < 0.02$).

In the future, the influence of surfactants and highly viscous droplets ($\lambda \gg 1$) on the excess velocity should be investigated to extend the model, since an excess velocity $u_{ex} < 1$ has not been included into the model so far. We suggest to include this function in the modeling of the wall film resistance. Especially local spatially resolved measurement techniques, e.g., µPIV measurements, should be appropriate for this task.

Author Contributions: Conceptualization, T.H. and U.M.; methodology, T.H., J.T. and U.M.; software, T.H., P.K. and U.M.; validation, T.H.; formal analysis, T.H. and U.M.; investigation, T.H.; resources, T.H.; data curation, T.H.; writing—original draft preparation, T.H.; writing—review and editing, U.M. and J.T.; visualization, T.H. and P.K.; and supervision, U.M. and J.T.

Funding: The contribution of P. Kemper has been supported by the German Research Foundation (DFG), Priority Program: "Reactive Bubbly Flows", SPP1740. The author gratefully acknowledges the financial support.

Conflicts of Interest: The authors declare no conflict of interest.

Abbreviations

The following abbreviations are used in this manuscript:

Acronyms/Abbreviations	
GA	Genetic Algorithm
PIV	Particle Image Velocimetry
PSA	Pattern Search Algorithm
Dimensionless Quantities	
λ	Viscosity ratio [−]
Ca	Capillary number [−]
Oh	Ohnesorge number [−]
Re	Reynolds number [−]
Greek Symbols	
β	geometric coefficient of resistance [−]
δ	wall film-thickness [µm]
η	dynamic viscosity [$\mathrm{Pa \cdot s}$]
Γ	control surface [-]
Ω	flow resistance [$\mathrm{Pa \cdot s \cdot m^{-3}}$]
ω	weight factor [−]
ψ	dimensionless equilibrium function [−]
ρ	density [$\mathrm{kg \cdot m^{-3}}$]
σ	interfacial tension [$\mathrm{N \cdot m^{-1}}$]
Roman Symbols	
A	cross-sectional area [$\mathrm{\mu m^2}$]
a	semi-minor axis of ellipse [µm]
ar	aspect ratio [−]
b	semi-major axis of ellipse [µm]
c	fitting coefficient [−]
D	hydraulic diameter [µm]

d	characteristic length [m]
H	channel height [μm]
h	droplet height [μm]
k	dimensionless ratio [−]
l	length [μm]
m	fitting coefficient [−]
n	fitting coefficient [−]
Q	volume flow rate [μl · min^{-1}]
R	radius [μm]
u	velocity [mm · s^{-1}]
W	channel width [μm]
x	cap length till gutter entrance [μm]
Superscripts	
$\mathcal{M}$	model calibration data
$meas$	measurement data
mod	model results
rel	relative
$stat$	stationary
Subscripts	
0	superficial/total/offset
β	concerning resistance factor
c	continuous phase
c, b	back droplet cap
c, f	front droplet cap
c, i	front or back droplet cap
cap	concerning droplet cap curvature
ch	channel
d	disperse/droplet phase
ex	excess scale concept
f	wall film
fb	from front to back
g	gutter
g, b	gutter at back of droplet
g, f	gutter at front of droplet
g, i	i-th droplet gutter
LP	Laplace pressure
$mean$	arithmetic mean
rel	relative value
s	slug
$slip$	slipping scale concept
Other Symbols	
$\overline{...}$	averaged value

Appendix A. Considerations for Mixed Terms

For consideration of the mixed terms in u_{ex}, rearranging Equation (18) leads to the equation

$$u_{ex} - u_{ex}\frac{A_f}{A_{ch}} = \frac{Q_g^{rel}}{Q_0} + \frac{A_f}{A_{ch}} \tag{A1}$$

Our measurements show in agreement with Jose and Cubaud [22] excess velocities with values $u_{ex} < 0.4$ for $Ca < 0.2$. Additionally, we can assume $\frac{A_f}{A_{ch}} < 0.005$, as shown in Figure A1. Thus, one can say $u_{ex}\frac{A_f}{A_{ch}} < 0.002$ and therefore it can be considered small of higher order and can be neglected.

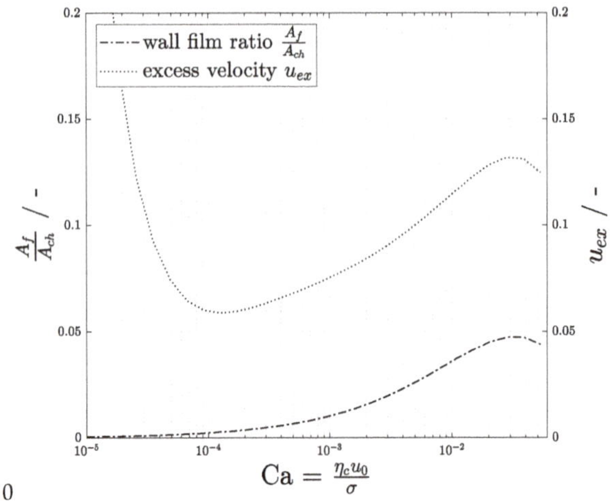

0

Figure A1. Dimensionless film-area and excess velocity for $\frac{l_d}{W} = 3$ and $\beta = 3.892$ for the proposed model.

The correlations, which were retrieved via a GA (Section 5), show a good agreement with the measurement data for the cap curvature from our last work and available literature data (Figure A2). Rising viscous forces indicated by a rising *Ca* number deform the droplet interface more strongly. At static conditions ($Ca \to 0$), the cap curvatures are in correspondence with Musterd et al. [45] roughly circular ($k_{c,f} = k_{c,b} \approx 1$). The back cap is compressed with respect to the main flow direction ($k_{c,b} < 1$) and the front cap elongated ($k_{c,f} > 1$) if viscous forces rise in comparison to the static case.

The size of the gutter radii rises with the increasing influence of the viscous forces, since the bypass flow in the gutter increases and needs to be accommodated by the gutters. The gutter entrance is always larger in diameter than the exit radius ($k_{g,f} > k_{g,b}$).

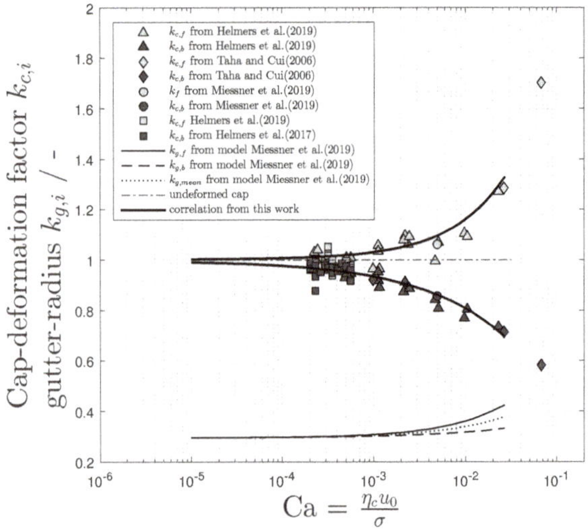

Figure A2. Measured values and retrieved correlation for the droplet cap deformation ratios $k_{c,f}$ and $k_{c,b}$ for different *Ca* numbers used in the proposed model. Additionally, the calculated dimensionless gutter-radii $k_{g,f}$, $k_{g,b}$ based on the model of Mießner et al. [24] are shown.

Appendix B. Criterion for Neglecting the Viscous Heating

Especially for microfluidic applications, flow phenomena such as viscous dissipation gain importance at specific scaling effects. Emerging from the energy dissipation, the temperature of the fluids can change (viscous heating) [55], as can relevant flow properties such as the viscosity change. To elaborate whether the viscous heating should be taken into account for the model, Morini [43] introduced a criterion, suggesting a bulk phase temperature increase of 1 °C as acceptable to still consider the viscosity unchanged. For adiabatic channels, they provided information of the critical *Re* depending on the hydraulic diameter. For d_{hyd} = 200 µm and rectangular channels, they stated $Re \approx 1000$.

Considering the validity range of the proposed model for *Ca* and *H*, this critical *Re* is reached under no circumstances using u_0 and A_{ch} from our measurements as properties.

When focusing on the gutters instead, local *Re* of

$$Re_g = \frac{\frac{Q_0}{A_g} D_g \rho_c}{\eta_c} < 16 \quad \text{(A2)}$$

are reached, if all four gutters are considered as one area equivalent circular tube with the diameter

$$D_g = \sqrt{4 \frac{A_g}{\pi}} \quad \text{(A3)}$$

In addition, these maximum values of $Re_g \approx 16$ lie below the critical $Re_c \approx 1000$ following Morini [43] for circular microchannels. Thus, we consider it valid to neglect the viscous heating in the applied range of *Ca* and *Re* of the proposed model.

References

1. Haase, S.; Murzin, D.Y.; Salmi, T. Review on hydrodynamics and mass transfer in minichannel wall reactors with gas–liquid Taylor flow. *Chem. Eng. Res. Des.* **2016**, *113*, 304–329. [CrossRef]
2. Sattari-Najafabadi, M.; Nasr Esfahany, M.; Wu, Z.; Sunden, B. Mass transfer between phases in microchannels: A review. *Chem. Eng. Process. Process Intensif.* **2018**, *127*, 213–237. [CrossRef]
3. Kralj, J.G.; Sahoo, H.R.; Jensen, K.F. Integrated continuous microfluidic liquid-liquid extraction. *Lab Chip* **2007**, *7*, 256–263. [CrossRef] [PubMed]
4. Kobayashi, J.; Mori, Y.; Kobayashi, S. Multiphase organic synthesis in microchannel reactors. *Chem. Asian J.* **2006**, *1*, 22–35. [CrossRef] [PubMed]
5. Wong, S.; Ward, M.; Wharton, C. Micro T-mixer as a rapid mixing micromixer. *Sens. Actuators B Chem.* **2004**, *100*, 359–379. [CrossRef]
6. Sun, B.; Jiang, J.; Shi, N.; Xu, W. Application of microfluidics technology in chemical engineering for enhanced safety. *Process Saf. Prog.* **2016**, *35*, 365–373. [CrossRef]
7. Chen, Y.; Li, P.; Huang, P.H.; Xie, Y.; Mai, J.D.; Wang, L.; Nguyen, N.T.; Huang, T.J. Rare cell isolation and analysis in microfluidics. *Lab Chip* **2014**, *14*, 626–645. [CrossRef]
8. Hosokawa, M.; Nishikawa, Y.; Kogawa, M.; Takeyama, H. Massively parallel whole genome amplification for single-cell sequencing using droplet microfluidics. *Sci. Rep.* **2017**, *7*, 5199. [CrossRef]
9. Jin, S.H.; Jung, J.H.; Jeong, S.G.; Kim, J.; Park, T.J.; Lee, C.S. Microfluidic dual loops reactor for conducting a multistep reaction. *Front. Chem. Sci. Eng.* **2018**, *12*, 239–246. [CrossRef]
10. Kück, U.D.; Mießner, U.; Aydin, M.; Thöming, J. Mixing time and mass transfer of rising bubbles in swarm turbulence. *Chem. Eng. Sci.* **2018**, *187*, 367–376. [CrossRef]
11. Mießner, U.; Kück, U.D.; Haase, K.; Kähler, C.J.; Fritsching, U.; Thöming, J. Experimental Assessment of an Innovative Device to Mimic Bubble Swarm Turbulence. *Chem. Eng. Technol.* **2017**, *40*, 1466–1474. [CrossRef]
12. Kreutzer, M.T.; Kapteijn, F.; Moulijn, J.A.; Heiszwolf, J.J. Multiphase monolith reactors: Chemical reaction engineering of segmented flow in microchannels. *Chem. Eng. Sci.* **2005**, *60*, 5895–5916. [CrossRef]

13. Van Steijn, V.; Kreutzer, M.; Kleijn, C. Velocity fluctuations of segmented flow in microchannels. *Chem. Eng. J.* **2008**, *135*, S159–S165. [CrossRef]
14. Abiev, R.; Svetlov, S.; Haase, S. Hydrodynamics and Mass Transfer of Gas-Liquid and Liquid-Liquid Taylor Flow in Microchannels. *Chem. Eng. Technol.* **2017**, *40*, 1985–1998. [CrossRef]
15. Gascon, J.; van Ommen, J.R.; Moulijn, J.A.; Kapteijn, F. Structuring catalyst and reactor—An inviting avenue to process intensification. *Catal. Sci. Technol.* **2015**, *5*, 807–817. [CrossRef]
16. Kreutzer, M.T.; Bakker, J.J.W.; Kapteijn, F.; Moulijn, J.A.; Verheijen, P.J.T. Scaling-up Multiphase Monolith Reactors: Linking Residence Time Distribution and Feed Maldistribution. *Ind. Eng. Chem. Res.* **2005**, *44*, 4898–4913. [CrossRef]
17. Huerre, A.; Miralles, V.; Jullien, M.C. Bubbles and foams in microfluidics. *Soft Matter* **2014**, *10*, 6888–6902. [CrossRef] [PubMed]
18. Fu, T.; Ma, Y. Bubble formation and breakup dynamics in microfluidic devices: A review. *Chem. Eng. Sci.* **2015**, *135*, 343–372. [CrossRef]
19. Ładosz, A.; von Rohr, P.R. Pressure drop of two-phase liquid-liquid slug flow in square microchannels. *Chem. Eng. Sci.* **2018**, *191*, 398–409. [CrossRef]
20. Schubert, M.; Kost, S.; Lange, R.; Salmi, T.; Haase, S.; Hampel, U. Maldistribution susceptibility of monolith reactors: Case study of glucose hydrogenation performance. *AIChE J.* **2016**, *62*, 4346–4364. [CrossRef]
21. Abiev, R. Analysis of local pressure gradient inversion and form of bubbles in Taylor flow in microchannels. *Chem. Eng. Sci.* **2017**, *174*, 403–412. [CrossRef]
22. Jose, B.M.; Cubaud, T. Formation and dynamics of partially wetting droplets in square microchannels. *RSC Adv.* **2014**, *4*, 14962–14970. [CrossRef]
23. Helmers, T. On the Excess Velocity of Taylor-Droplets in Square Microchannels. Ph.D. Thesis, University of Bremen, Bremen, Germany, 2019.
24. Mießner, U.; Helmers, T.; Lindken, R.; Westerweel, J. An analytical interface shape approximation of microscopic Taylor flows. *Exp. Fluids* **2019**, *60*, 75. [CrossRef]
25. Helmers, T.; Kemper, P.; Thöming, J.; Mießner, U. Determining the flow-related cap deformation of Taylor droplets at low Ca numbers using ensemble-averaged high-speed images. *Exp. Fluids* **2019**, *60*, 66. [CrossRef]
26. Direito, F.; Campos, J.; Miranda, J.M. A Taylor drop rising in a liquid co-current flow. *Int. J. Multiph. Flow* **2017**, *96*, 134–143. [CrossRef]
27. Shams, M.; Raeini, A.Q.; Blunt, M.J.; Bijeljic, B. A study to investigate viscous coupling effects on the hydraulic conductance of fluid layers in two-phase flow at the pore level. *J. Colloid Interface Sci.* **2018**, *522*, 299–310. [CrossRef]
28. Ohnesorge, W.V. Die Bildung von Tropfen an Düsen und die Auflösung flüssiger Strahlen. *J. Appl. Math. Mech.* **1936**, *16*, 355–358. [CrossRef]
29. Wong, H.; Radke, C.J.; Morris, S. The motion of long bubbles in polygonal capillaries. Part 2. Drag, fluid pressure and fluid flow. *J. Fluid Mech.* **1995**, *292*, 95. [CrossRef]
30. Rao, S.S.; Wong, H. The motion of long drops in rectangular microchannels at low capillary numbers. *J. Fluid Mech.* **2018**, *852*, 60–104. [CrossRef]
31. Jakiela, S.; Makulska, S.; Korczyk, P.M.; Garstecki, P. Speed of flow of individual droplets in microfluidic channels as a function of the capillary number, volume of droplets and contrast of viscosities. *Lab Chip* **2011**, *11*, 3603–3608. [CrossRef]
32. Kalantarifard, A.; Saateh, A.; Elbuken, C. Label-Free Sensing in Microdroplet-Based Microfluidic Systems. *Chemosensors* **2018**, *6*, 23. [CrossRef]
33. Liu, H.; Vandu, C.O.; Krishna, R. Hydrodynamics of Taylor Flow in Vertical Capillaries: Flow Regimes, Bubble Rise Velocity, Liquid Slug Length, and Pressure Drop. *Ind. Eng. Chem. Res.* **2005**, *44*, 4884–4897. [CrossRef]
34. Howard, J.A.; Walsh, P.A. Review and extensions to film thickness and relative bubble drift velocity prediction methods in laminar Taylor or slug flows. *Int. J. Multiph. Flow* **2013**, *55*, 32–42. [CrossRef]
35. Angeli, P.; Gavriilidis, A. Hydrodynamics of Taylor flow in small channels: A Review. *Proc. Inst. Mech. Eng. Part C* **2008**, *222*, 737–751. [CrossRef]
36. Abadie, T.; Aubin, J.; Legendre, D.; Xuereb, C. Hydrodynamics of gas–liquid Taylor flow in rectangular microchannels. *Microfluid. Nanofluid.* **2012**, *12*, 355–369. [CrossRef]
37. Bretherton, F.P. The motion of long bubbles in tubes. *J. Fluid Mech.* **1961**, *10*, 166–188. [CrossRef]

38. Rocha, L.; Miranda, J.; Campos, J. Wide Range Simulation Study of Taylor Bubbles in Circular Milli and Microchannels. *Micromachines* **2017**, *8*, 154. [CrossRef]
39. Hangos, K.M.; Cameron, I.T. (Eds.) *Process Modelling and Model Analysis*; Academic Press: San Diego, CA, USA, 2001; Volume 4.
40. Han, Y.; Shikazono, N. Measurement of liquid film thickness in micro square channel. *Int. J. Multiph. Flow* **2009**, *35*, 896–903. [CrossRef]
41. Kreutzer, M.T.; Shah, M.S.; Parthiban, P.; Khan, S.A. Evolution of nonconformal Landau-Levich-Bretherton films of partially wetting liquids. *Phys. Rev. Fluids* **2018**, *3*, 42. [CrossRef]
42. Khodaparast, S.; Atasi, O.; Deblais, A.; Scheid, B.; Stone, H.A. Dewetting of Thin Liquid Films Surrounding Air Bubbles in Microchannels. *Langmuir* **2018**, *34*, 1363–1370. [CrossRef]
43. Morini, G.L. Viscous heating in liquid flows in micro-channels. *Int. J. Heat Mass Transf.* **2005**, *48*, 3637–3647. [CrossRef]
44. Abate, A.R.; Mary, P.; van Steijn, V.; Weitz, D.A. Experimental validation of plugging during drop formation in a T-junction. *Lab Chip* **2012**, *12*, 1516–1521. [CrossRef]
45. Musterd, M.; van Steijn, V.; Kleijn, C.R.; Kreutzer, M.T. Calculating the volume of elongated bubbles and droplets in microchannels from a top view image. *RSC Adv.* **2015**, *5*, 16042–16049. [CrossRef]
46. Bruus, H. *Theoretical Microfluidics*; Oxford University Press: Oxford, UK, 2008; Volume 18, p. 76.
47. Ransohoff, T.; Radke, C. Laminar flow of a wetting liquid along the corners of a predominantly gas-occupied noncircular pore. *J. Colloid Interface Sci.* **1988**, *121*, 392–401. [CrossRef]
48. Kreutzer, M.T.; Kapteijn, F.; Moulijn, J.A.; Kleijn, C.R.; Heiszwolf, J.J. Inertial and interfacial effects on pressure drop of Taylor flow in capillaries. *AIChE J.* **2005**, *51*, 2428–2440. [CrossRef]
49. Whitley, D. A genetic algorithm tutorial. *Stat. Comput.* **1994**, *4*. [CrossRef]
50. Davey, K.R. Latin Hypercube Sampling and Pattern Search in Magnetic Field Optimization Problems. *IEEE Trans. Magn.* **2008**, *44*, 974–977. [CrossRef]
51. Helmers, T.; Thöming, J.; Mießner, U. Retrieving accurate temporal and spatial information about Taylor slug flows from non-invasive NIR photometry measurements. *Exp. Fluids* **2017**, *58*, 66. [CrossRef]
52. Fuerstman, M.J.; Lai, A.; Thurlow, M.E.; Shevkoplyas, S.S.; Stone, H.A.; Whitesides, G.M. The pressure drop along rectangular microchannels containing bubbles. *Lab Chip* **2007**, *7*, 1479–1489. [CrossRef] [PubMed]
53. Jakiela, S.; Korczyk, P.M.; Makulska, S.; Cybulski, O.; Garstecki, P. Discontinuous transition in a laminar fluid flow: A change of flow topology inside a droplet moving in a micron-size channel. *Phys. Rev. Lett.* **2012**, *108*, 134501. [CrossRef] [PubMed]
54. Kumari, S.; Kumar, N.; Gupta, R. Flow and heat transfer in slug flow in microchannels: Effect of bubble volume. *Int. J. Heat Mass Transf.* **2019**, *129*, 812–826. [CrossRef]
55. Celata, G.P.; Morini, G.L.; Marconi, V.; McPhail, S.J.; Zummo, G. Using viscous heating to determine the friction factor in microchannels—An experimental validation. *Exp. Therm. Fluid Sci.* **2006**, *30*, 725–731. [CrossRef]

fluids

MDPI

Article

Dynamics of an Ellipse-Shaped Meniscus on a Substrate-Supported Drop under an Electric Field

Philip Zaleski and Shahriar Afkhami *

Department of Mathematical Sciences, New Jersey Institute of Technology, Newark, NJ 07102, USA; pz85@njit.edu
* Correspondence: shahriar.afkhami@njit.edu

Received: 1 October 2019; Accepted: 26 November 2019; Published: 29 November 2019

Abstract: The behavior of a conducting droplet and a dielectric droplet placed under an electric potential is analyzed. Expressions for drop height based on electrode separation and the applied voltage are found, and problem parameters associated with breakup and droplet ejection are classified. Similar to previous theoretical work, the droplet interface is restricted to an ellipse shape. However, contrary to previous work, the added complexity of the boundary condition at the electrode is taken into account. To gain insight into this problem, a two-dimensional droplet is addressed. This allows for conformal maps to be used to solve for the potential surrounding the drop, which gives the total upward electrical force on the drop that is then balanced by surface tension and gravitational forces. For the conducting case, the maximum droplet height is attained when the distance between the electrode and the drop becomes sufficiently large, in which case, the droplet can stably grow to about 2.31 times its initial height before instabilities occur. In the dielectric case, hysteresis can occur for certain values of electrode separation and relative permittivity.

Keywords: electrified fluids; conformal map; Taylor cone

1. Introduction

Electrified fluids appear in a wide variety of applications, such as inkjet printing [1–3], electrospray ionization/mass spectrometry [4,5], electrospinning [6,7], focused ion beam (FIB) technology [8,9], and nanotechnology [10]. However, an early motivation for Rayleigh and Taylor to study the behavior of droplets in electric fields came from nature, as electrified fluids play an important role in producing thunderstorms [11,12]. Many early theoretical papers by Taylor [11], and others, primarily focused on droplets suspended in uniform fields [13,14]. To analyze this problem, Taylor assumed that the interface was an ellipsoidal shape, derived a two-point approximation satisfying the Young-Laplace's equation at the poles and equator of the ellipsoid, which resulted in an expression relating droplet height to the strength of the surrounding field. After Taylor's original work on conducting drops in uniform fields, Miksis [15] extended Taylor's approach to dielectric drops, and also performed numerical work which showed that hysteresis or bistability could occur for certain values of relative permittivity.

However, in many applications and experiments, an electric potential is placed directly above the droplet, and the electric field surrounding the drop is not uniform. This creates a new problem, which is sketched in Figure 1. Taylor and McEwan theoretically analyzed this problem, but instead of assuming an ellipsoidal shape and solving for the surrounding potential, they assumed that the interface was horizontal at its poles and guessed a potential that satisfied the upper boundary condition at the electrode [16]. Since then, Corson et al. [17,18] theoretically analyzed a conducting drop in the limiting case where the distance between the substrate and the electrode was large, and obtained asymptotic results which predicted drop height for small surrounding electric fields.

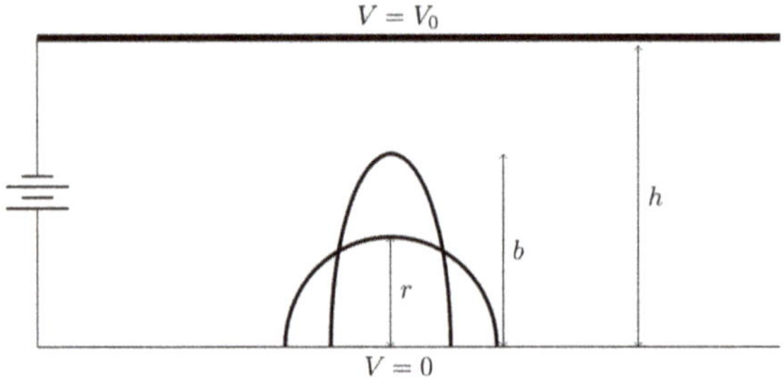

Figure 1. Two-dimensional sketch of the problem. The initial radius of the cylindrical drop is given by r, the distance between the substrate and the electrode by h, the applied voltage by V_0, and the new height of the drop with the applied voltage by b.

When a sessile drop on a substrate is placed under an electric potential difference between the substrate and an electrode above the drop, for sufficiently large potentials (where ellipsoidal shapes are no longer stable), the drop forms a conical shape known as the Taylor cone [11]. In [6], Yarin et al. showed that the Taylor cone which is a specific self-similar solution is not unique and that there could also exist nonself-similar solutions that do not admit a Taylor cone. The theory and numerical simulations of the jetting of the Taylor cone has been described in some recent papers, e.g., [19–21]. Recently, a study conducted by Elele et al. [22] sparked new interests in this problem, as it experimentally analyzed a dynamic Taylor cone and found three possible modes that the droplet could form depending on the applied voltage. The first mode happened for small voltages when the droplet took on an ellipsoidal shape and steadily rose as the applied voltage increased. Another mode happened for the largest applied voltages where strong instabilities such as droplet ejection arose. An intermediate mode happened for moderate voltages where the droplet would form a pointed protrusion and periodically rose to touch the electrode, and in this mode, a universal self symmetry independent of the applied voltage was observed [22]. This work served as an early motivation for us to study the behavior of an electrified droplet.

In this paper, we analyze a perfectly conducting drop and a dielectric drop to mathematically understand why droplet behavior in both cases is similar for large values of permittivity but drastically different for small values of permittivity, and why bistability only happens for in-between values of permittivity. We take the boundary condition at the electrode into account and derive formulas that signify when instabilities such as droplet ejection or droplet beak-up occur. To achieve this, we address the two-dimensional version of this problem. We assume that the droplet takes on an ellipse shape, which allows us to approximate the potential surrounding the droplet and obtain the total upward electrical force on the droplet that is then balanced by surface tension and gravitational forces. These assumptions describe a problem that is different from the one that has been experimentally studied in [12,17,22,23], and thus, our exact values for droplet height and applied voltage that are associated with droplet stability deviate slightly from those observed in experiments. However, our approach reasonably predicts the overall behavior of the drop and gives further insight into how droplet behavior changes as a function of the permittivity.

2. Conducting Drop

For the case of a perfectly conducting drop, we have that the potential V_E surrounding the drop satisfying the Laplace's equation,

$$\Delta V_E = 0, \tag{1}$$

with a Dirichlet boundary condition at the electrode,

$$V_E(0,h) = V_0, \tag{2}$$

a homogeneous Dirichlet boundary condition at the droplet interface, which we parameterize by t,

$$V_E(a\cos t, b\sin t) = 0, \tag{3}$$

a homogeneous Neumann boundary condition far away from the interface,

$$\frac{\partial V_E}{\partial x}(x_0, y) = 0, \tag{4}$$

and finally, we take advantage of the symmetry about the major axis of the ellipse and only analyze the part of the interface which corresponds to $0 < t < \frac{\pi}{2}$, which gives one extra homogeneous Neumann boundary condition on the y-axis,

$$\frac{\partial V_E}{\partial x}(0, y) = 0. \tag{5}$$

To solve for V_E, we use the conformal map defined by,

$$f(z) = \frac{az + \sqrt{(bz)^2 - (a+b)^2 b^2}}{a+b}, \tag{6}$$

which maps a rectangle with height $\hat{h}$ to a region that approximates the domain that we are working on (i.e., it maps the region on the right in Figure 2 to the region on the left), where

$$\hat{h} = \frac{ah - \sqrt{b^2h^2 + a^2b^2 - b^4}}{a-b}. \tag{7}$$

Thus, we have that the inverse of this map defined by,

$$g(x+iy) = \frac{1}{a-b}\left(ax - \sqrt{\frac{\sqrt{(b^2(x^2-y^2) - a^2b^2 + b^4)^2 + 4b^4x^2y^2} + b^2(x^2-y^2) - a^2b^2 + b^4}{2}} + i\left(a\,y - \sqrt{\frac{\sqrt{(b^2(x^2-y^2) - a^2b^2 + b^4)^2 + 4b^4x^2y^2} - b^2(x^2-y^2) + a^2b^2 - b^4}{2}}\right)\right) \tag{8}$$

will map a region which approximates our domain into a simple rectangle with height $\hat{h}$, which gives us the approximate surrounding potential,

$$V_E(x,y) = \frac{V_0}{\hat{h}(a-b)}\left(a\,y - \sqrt{\frac{\sqrt{(b^2(x^2-y^2) - a^2b^2 + b^4)^2 + 4b^4x^2y^2} - b^2(x^2-y^2) + a^2b^2 - b^4}{2}}\right). \tag{9}$$

This gives the potential field shown in Figure 3 that is zero at the interface, and approximately linear far away from the interface.

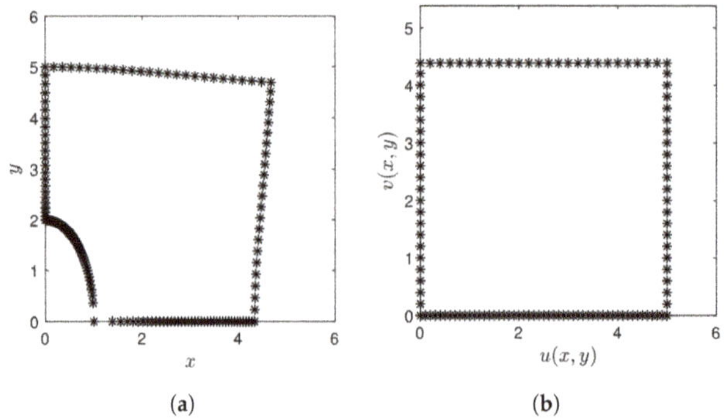

Figure 2. The map given in Equation (6) maps from (**b**) to (**a**), and the inverse map given in Equation (8) maps in the opposite direction.

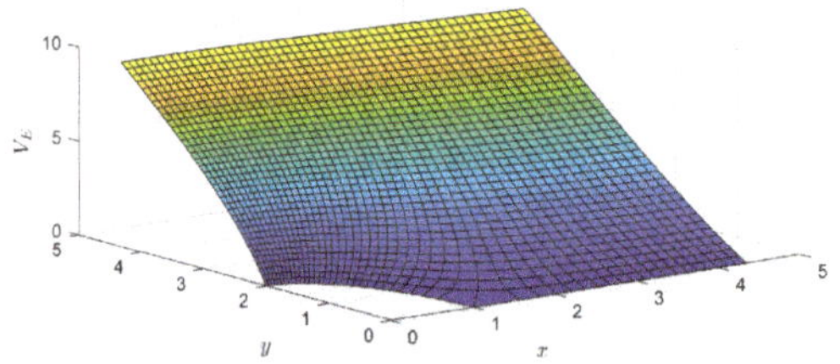

Figure 3. Example of the plot of the potential given in Equation (9) for $b = 2$, $a = 1$, and $h = 5$.

While the potential given above is not an exact solution to Equation (1) with the corresponding boundary conditions, as the boundary condition at the electrode is slightly warped, for most values of b and h the relative error at the interface is small. For instance, Figure 4a plots Error_V for varying values of b where Error_V is defined as

$$\text{Error}_V = \frac{||V_E - V_n||_1}{||V_E||_1}. \tag{10}$$

Here V_n stands for a numerical finite element solution to Laplace's equation on the exact domain, and the L^1-norm is approximated by the function values at the grid points of the scheme. Furthermore, the error in the electric field at the interface which we define as

$$\text{Error}_E = \frac{||(\nabla V_E \cdot n - \nabla V_n \cdot n)_{int}||_1}{||(\nabla V_E \cdot n)_{int}||_1}, \tag{11}$$

is plotted in Figure 4b for varying values of b (where $(\cdot)_{int}$ stands for $\cdot$ evaluated the interface). Both Figure 4a,b indicate that our analytical solution breaks down as b approaches h, as in this regime the top and right boundaries of our conformal mapped domain deviate from that of straight lines (see Figure 4c–e).

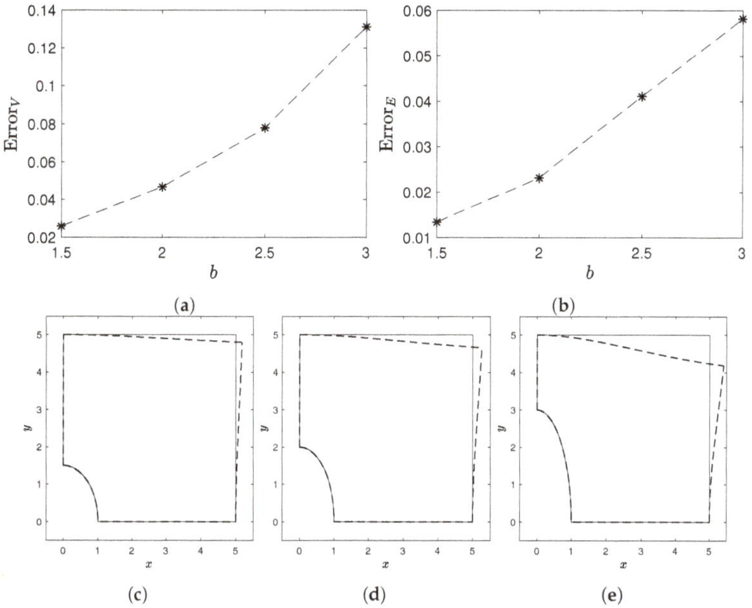

Figure 4. (**a**) The relative error between our analytical potential given in Equation (9) and a finite element solution for the potential on the exact domain, which we define as Error$_V$ (see Equation (10)) is potted for varying values of b. (**b**) The relative error in the electric field strength at the droplet interface given by our analytical solution (see Equation (12)) and a finite element solution on the exact domain, which we define as Error$_E$ (see Equation (11)) is plotted for varying values of b. (**c–e**) The exact domain (—–) is plotted against our approximate conformal mapped domain (- - -) for $b = 1.5$, $b = 2$, and $b = 3$. (**a–e**) We fix h to 5 and a to 1.

3. Total Force on the Conducting Drop

The potential in Equation (9) provides the surrounding electric field,

$$\vec{E} = -\nabla V_E = \frac{-V_0}{\hat{h}(a-b)} \begin{pmatrix} E_x \\ E_y \end{pmatrix}, \tag{12}$$

where

$$E_x = \frac{bx\left(a^2-b^2-x^2-y^2+\sqrt{4x^2y^2+(a^2-b^2-x^2+y^2)^2}\right)}{\sqrt{2}\sqrt{(4x^2y^2+(a^2-b^2-x^2+y^2)^2)(a^2-b^2-x^2+y^2+\sqrt{4x^2y^2+(a^2-b^2-x^2+y^2)^2})}}, \tag{13}$$

and,

$$E_y = a - \frac{by(a^2-b^2+x^2+y^2+\sqrt{4x^2y^2+(a^2-b^2-x^2+y^2)^2})}{\sqrt{2}\sqrt{(4x^2y^2+(a^2-b^2-x^2+y^2)^2)(a^2-b^2-x^2+y^2+\sqrt{4x^2y^2+(a^2-b^2-x^2+y^2)^2})}}. \tag{14}$$

This gives a total electrical force per-unit area that is normal to the interface with magnitude given by,

$$f_{E_n} = \frac{1}{2}\epsilon_2 E_{int}^2, \tag{15}$$

where E_{int} stands for $|\vec{E}|$ evaluated at the interface, and ϵ_2 is the permittivity of the medium surrounding the droplet [24,25]. In our case, this medium is air which means that $\epsilon_2 \approx \epsilon_0$. Setting $x = a\cos t$ and $y = b\sin t$, we evaluate Equation (15) at the interface to get that

$$f_{E_n} = \frac{\epsilon_2 V_0^2}{\hat{h}^2}\frac{(a+b)^2\sin^2 t}{2(b^2\cos^2 t + a^2\sin^2 t)}, \tag{16}$$

which gives the force distribution shown in Figure 5.

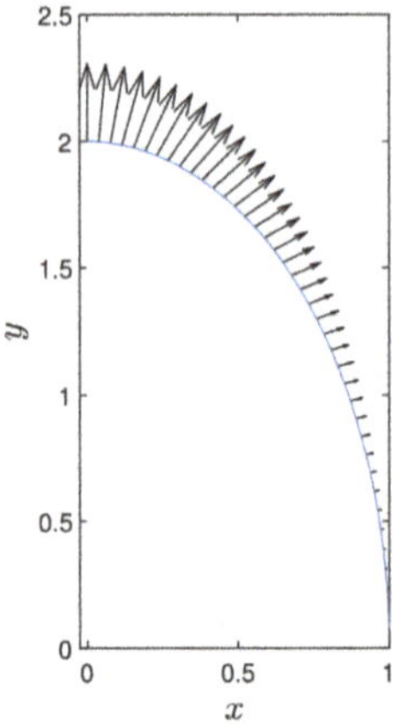

Figure 5. Electrical force distribution around a conducting drop.

Integrating the expression given in Equation (16) and accounting for surface tension, pressure (we neglect the effect that the non-constant curvature of the ellipse has on the Laplace pressure and assume that the internal pressure is constant as in [11]), and gravitational forces dictates that the total upward force per-unit length on our cylindrical drop is,

$$F_{Total} = -\rho\frac{\pi r^2 g}{2} + \gamma\left(\frac{2a}{r} - 2\right) + \frac{\epsilon_0 V_0^2}{\hat{h}^2}\left(\frac{a(a+b)}{a-b} + \frac{b^2\sqrt{b+a}}{(b-a)^{\frac{3}{2}}}\arctan\left(\frac{\sqrt{b^2-a^2}}{a}\right)\right). \quad (17)$$

Here r is the radius of the semi-circular drop, ρ is the density of the fluid, γ is the surface tension constant, and g is the acceleration due to gravity. Assuming incompressibility of the liquid gives us $a = \frac{r^2}{b}$, which allows us to express the total upward force per-unit length on the drop in terms of b.

Finally, we pick our characteristic length scale to be r and nondimensionalize Equation (17) to get that

$$F = -Bo_g + \tfrac{1}{B} - 1 + Bo_e\left(\tfrac{R^2}{(R-B\sqrt{R^2B^2+1-B^4})^2}\left(\tfrac{1-B^4}{B} + B^3\sqrt{B^4-1}\arctan\left(\sqrt{B^4-1}\right)\right)\right), \quad (18)$$

where our new variables are

$$F = \frac{F_{Total}}{2\gamma}, \quad (19)$$

and

$$B = \frac{b}{r}. \quad (20)$$

Our problem parameters are,

$$Bo_g = \rho\frac{\pi r^2 g}{4\gamma}, \quad (21)$$

$$Bo_e = \frac{\epsilon_0 V_0^2 r}{2\gamma h^2}, \quad (22)$$

and

$$R = \frac{h}{r}. \quad (23)$$

4. Results

Setting the total upward force on the droplet given in Equation (18) equal to zero gives us Bo_e in terms of Bo_g, R, and B. For simplicity, we set $Bo_g = 0$, and plot Bo_e in terms of B for various values of electrode separation R. Similar to earlier analytical results by Taylor [11] and Miksis [15] for droplets in uniform fields and numerical results for the full three dimensional problem in [15,26,27], our method gives one stable branch of fixed points (solid line) and one unstable branch (dotted line). From Figure 6 which plots $\log(Bo_e + 1)$ vs $\log B$ for varying values of R, we can see that our formula for two-dimensional drops, which takes into account electrode separation allows for larger drop heights and applied voltages than previous theoretical results for ellipsoidal drops by Miksis [15]. However, in Figure 6 we can also see that for small voltages our two dimensional approximation and Miksis's [15] three dimensional approximation seem to agree well. Furthermore, from Figure 6, we have that for each value of R there is a certain threshold value of Bo_e and a corresponding value of B which identifies the maximum stable height that the droplet can grow to. These threshold values are plotted in Figure 7, showing that a two-dimensional drop can only grow to 2.31 its initial height before instabilities arise and that this maximum height happens when $Bo_e = 0.14$, and in the limiting case where $R \to \infty$ or identically $h \to \infty$. This maximum value of $B = 2.31$ deviates from Taylor's analytical results in three dimensions which predict a maximum value of $B = 1.38$ [11], and Taylor's experiential results with thin films which predict a maximum value of $B = 1.48$ [12]. However, our results are in agreement with experimental values in [22,23]. For instance, experiments by Macky [23] show that a free-falling drop in an electric field can grow to anywhere between 1.7 and 2.2 its original length before instabilities occur. Experiments done by Inculet and Kromann [28] on water droplets doped with alcohol and suspended in oil show that the droplet can stably grow to 2.15 its initial height. Most recently, experiments done by Elele et al. [22] on electrified droplets on the International Space Station show that a droplet can stably grow to 2.61 its original length. However, it is also important to note that in many of these experiments, the curvature at the tip of the droplet may be starting to deviate from that of an ellipse. Furthermore, this deviation from an ellipse shape might also be the reason why experiments give a voltage threshold that is higher than the one given by our method. For example, experiments carried out by Elele et al. [22] on a drop of slightly conducting water on Earth gravity with $r = 1.68$ mm and an electrode separation of $h = 2.10$ mm, found that the maximum voltage that could be applied before spontaneous shoot-up occurred was $V_0 = 1.61$ kV. However, our method predicts that for these problem parameters, the maximum voltage that can be applied before an ellipse is no longer a stable solution is $V_0 = 1.19$ kV. Nevertheless, in this case, gravitational forces are strong in comparison to the electrical forces, and because of this, electrode separation is quite small which means that the error in our calculated potential could also be having an effect.

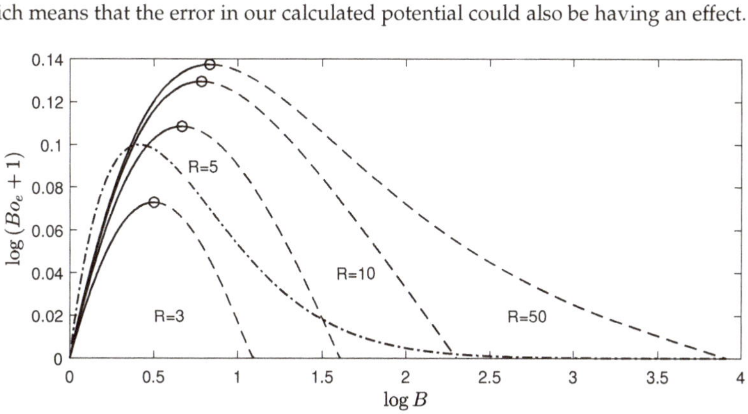

Figure 6. Electrical bond number Bo_e is plotted against B on a log- log scale for $R = 3$, $R = 5$, $R = 10$, and $R = 50$ (——, - - -), and values of Bo_e obtained from Miksis's [15] generalization of Taylor's two point method for 3-dimensional ellipsoidal droplets in uniform electric fields is also plotted (-··-·).

Here we can see that the stable branch of fixed points approaches a fixed curve as $R \to \infty$, while the unstable branch of fixed points reaches all the way to R, which corresponds to unstable drop heights that approach the height of the electrode.

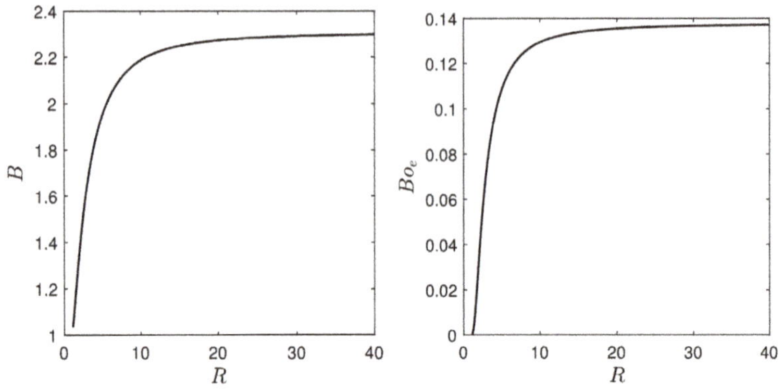

(**a**) (**b**)

Figure 7. (**a**) For each value of electrode separation (R), the maximum height that the droplet can stably achieve is plotted. (**b**) For each value of electrode separation (R), the applied potential (Bo_e) associated with this maximum height is plotted.

5. Dielectric Drop

For the case of a dielectric drop, the boundary condition at the interface changes; instead of a Dirichlet boundary condition, we have a Neumann boundary condition at the interface, which specifies the jump in permittivity and is given by

$$\epsilon_2 \nabla V_2(x(t), y(t)) \cdot \vec{n} - \epsilon_1 \nabla V_1(x(t), y(t)) \cdot \vec{n} = 0. \tag{24}$$

Here $\vec{n}$ is the unit normal vector of the interface, ϵ_1 is the permittivity of the fluid, V_2 is the electric potential outside the droplet, and V_1 is the electric potential inside the droplet. Note that this boundary condition assumes that the fluid is a perfect dielectric and that no free charges build up at the interface [25].

Because we consider the potential inside the droplet, we cannot use the map given in Equation (8), so instead we use the map given by

$$f(z) = i\sqrt{b^2 - a^2}\cos(z) = \sqrt{b^2 - a^2}(\sin(x)\sinh(y) + i\cos(x)\cosh(y)), \tag{25}$$

which maps the rectangle with width $\frac{\pi}{2}$ and height arctanh $(\frac{a}{b})$ to the region enclosed by the quarter of an Ellipse with major axis b and minor axis a. Furthermore, the horizontal line segment given by,

$$z = x + iH, \tag{26}$$

where $0 < x < \frac{\pi}{2}$, and

$$H = \cosh^{-1}\left(\frac{h}{\sqrt{b^2 - a^2}}\right), \tag{27}$$

gets mapped to the region enclosed by the quarter of an ellipse with major axis h and minor axis $\sqrt{h^2 - b^2 + a^2}$ (see Figure 8). From here, we can see that for most values of b, the width of this outer ellipse will be much bigger than a.

Thus, the inverse of this map, given by

$$w = f^{-1}(z) = \cos^{-1}\left(\frac{y - ix}{\sqrt{b^2 - a^2}}\right), \tag{28}$$

will allow us to solve the desired equations on an elliptical annulus; however, this requires us to define a boundary condition on the outer ellipse. To ensure that this boundary condition is physically relevant and approximates the boundary condition at the electrode, we define V_2 on this outer ellipse to be equal to our conducting potential V_E given in Equation (9). This turns out to be a good approximation for the boundary condition at the electrode, as we find that the error for our potential in the dielectric case is bounded by the error in the conducting case (results not show here).

Because we are only working in the first quadrant, we can define a branch of $f^{-1}(z) = X(x,y) + iY(x,y)$ that is analytic everywhere on the elliptical annulus except at the point $x = 0$, $y = \sqrt{b^2 - a^2}$. This branch of $f^{-1}(z) = X(x,y) + iY(x,y)$ is defined by the branch of arccos z given by,

$$\begin{aligned}
&\arccos(x + iy) = \\
&\operatorname{arctan2}\left(y + \sqrt{\frac{\sqrt{(x^2 - y^2 - 1)^2 + 4x^2y^2} - x^2 + y^2 + 1}{2}}, x - \sqrt{\frac{\sqrt{(x^2 - y^2 - 1)^2 + 4x^2y^2} + x^2 - y^2 - 1}{2}}\right) \\
&\qquad - \frac{i}{2}\ln\left(x^2 + y^2 + \sqrt{(x^2 - y^2 - 1)^2 + 4x^2y^2}\right. \\
&+\sqrt{2}\left(-x\sqrt{\sqrt{(x^2 - y^2 - 1)^2 + 4x^2y^2} + x^2 - y^2 - 1} + y\sqrt{\sqrt{(x^2 - y^2 - 1)^2 + 4x^2y^2} - x^2 + y^2 + 1}\right)\Bigg).
\end{aligned} \tag{29}$$

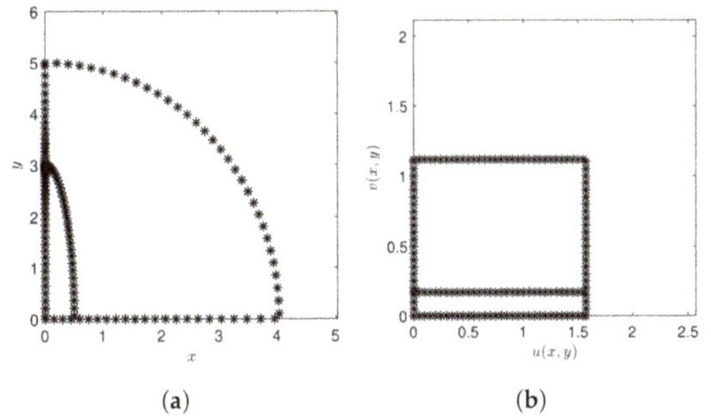

Figure 8. Example of the map given in Equation (25) and the branch of its inverse define by Equations (28) and (29) for $b = 3$, $a = \frac{1}{2}$, and $h = 5$. Here $f(z)$ maps from (**b**) to (**a**).

Using the fact that conformal maps preserve both the Laplace's equation and homogeneous Neumann boundary conditions, we arrive at,

$$V_1(x,y) = \frac{\epsilon_2 V_0 (b^2 - a^2) y}{\epsilon_1 (ab\sqrt{h^2 - b^2 + a^2} - a^2 h) + \epsilon_2 (hb^2 - ab\sqrt{h^2 - b^2 + a^2})}, \tag{30}$$

and,

$$\begin{aligned}
V_2(x,y) = \frac{V_0}{h}y + \\
\frac{V_0(\epsilon_1 - \epsilon_2)a\, b(b^2 - a^2)}{h(\epsilon_1(ab\sqrt{h^2 - b^2 + a^2} - a^2 h) + \epsilon_2(hb^2 - ab\sqrt{h^2 - b^2 + a^2}))}\left(\frac{h}{\sqrt{b^2 - a^2}}\cos(X)\sinh(Y) - \frac{\sqrt{h^2 - b^2 + a^2}y}{b^2 - a^2}\right),
\end{aligned} \tag{31}$$

where X and Y are given by a branch of the inverse transformation. Here we notice that the potential inside the droplet is linear, which implies that the field in that region is uniform, and we can also see that the potential inside the droplet approaches zero as $\epsilon_1 \to \infty$, which implies that in the limiting case where the permittivity of the drop goes to infinity V_1 goes to zero and V_2 goes to V_E (the potential surrounding the conducting drop). Additionally, the example plot (Figure 9) of the potential given in Equations (30) and (31) shows that while the potential inside the droplet is linear, the potential directly outside the drop is nonlinear and approaches a linear function for values of x, and y that are large.

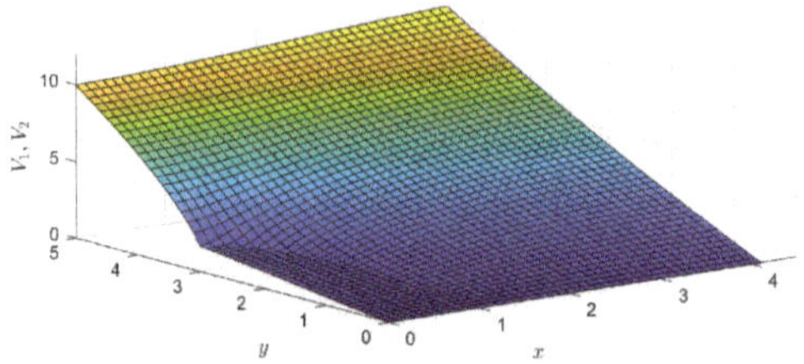

Figure 9. Example plot of the potential for the dielectric case when $b = 3$, $a = 1$, $h = 5$, and $V_0 = 10$.

6. Total Force on the Dielectric Drop

Equation (30) gives the vertical electric field inside the droplet, defined by,

$$E_1 = \frac{-\epsilon_2 V_0 (b^2 - a^2)}{\epsilon_1 (ab\sqrt{h^2 - b^2 + a^2} - a^2 h) + \epsilon_2 (hb^2 - ab\sqrt{h^2 - b^2 + a^2})} \begin{pmatrix} 0 \\ 1 \end{pmatrix}. \tag{32}$$

Equation (31) gives an electric field outside of the drop that has an x-component of,

$$E_{2x} = -\frac{V_0(\epsilon_1 - \epsilon_2) a\, b (b^2 - a^2)}{\epsilon_1 (ab\sqrt{h^2 - b^2 + a^2} - a^2 h) + \epsilon_2 (hb^2 - ab\sqrt{h^2 - b^2 + a^2})} \frac{\sin t \cos t}{a^2 \sin^2 t + b^2 \cos^2 t}, \tag{33}$$

and a y-component of,

$$E_{2y} = -\frac{V_0}{h} - \frac{V_0(\epsilon_1 - \epsilon_2) a\, b}{h(\epsilon_1 (ab\sqrt{h^2 - b^2 + a^2} - a^2 h) + \epsilon_2 (hb^2 - ab\sqrt{h^2 - b^2 + a^2}))} \left(\frac{abh}{(a^2 \sin^2 t + b^2 \cos^2 t)} - \sqrt{h^2 - b^2 + a^2} \right). \tag{34}$$

This allows to calculate the total force exerted at interface, yielding

$$f_n(t) = [\epsilon E_n] - \frac{1}{2}[\epsilon E^2], \tag{35}$$

where $f_n(t)$ is the normal electrical force per-unit area at the interface (note that since there are no free charges, the electrical force will be in the interface normal direction), and $[\cdot]$ stands for the jump in $\cdot$ at the interface [24,25]. The expression in Equation (35) gives the force distribution shown in Figure 10. From here, we observe that the electrical force distribution is almost uniform for small values of relative permittivity.

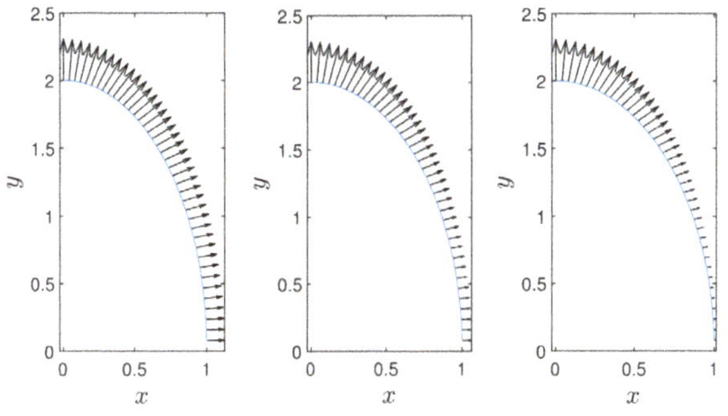

(**a**) (**b**) (**c**)

Figure 10. Electrical force distribution around the dielectric drop for (**a**) $\epsilon = 5$, (**b**) $\epsilon = 10$, and (**c**) $\epsilon = 50$.

By integrating Equation (35) from 0 to $\frac{\pi}{2}$, accounting for surface tension and gravitational forces, and non-dimensionalizing, we get that the total dimensionless force (see Equation (19)) on the droplet is,

$$F = -Bo_g + \frac{1}{B} - 1 + R^2 Bo_e \Bigg(\frac{2B}{\sqrt{B^4-1}} \Big(\frac{B^2(B^2\hat{C_1}(\epsilon-1)+\hat{C_2})^2}{B^4-1} - \epsilon\hat{C_1}^2(B^2 - \frac{1}{B^2}) - \frac{\hat{C_1}^2}{2}(\epsilon-1)^2(B^2+\frac{1}{B^2}) - \hat{C_1}\hat{C_2}(\epsilon-1) \Big) \Big(2\arctan\left(\sqrt{B^4-1}\right) - \arctan\left(\sqrt{B^4-1}-B^2\right) - \arctan\left(\sqrt{B^4-1}+B^2\right) \Big) - \frac{2(\hat{C_1}B^2(\epsilon-1)+\hat{C_2})^2}{B(B^4-1)} + \frac{2\epsilon\hat{C_1}^2}{B^3}(B^2-1/B^2) + \frac{\hat{C_1}^2}{B}(\epsilon-1)^2 - \frac{\hat{C_2}^2}{B} + \epsilon\hat{C_1}^2\frac{1}{B}(B^2-\frac{1}{B^2})^2 \Bigg), \quad (36)$$

where,

$$\hat{C_1} = \frac{1}{\epsilon\left(\sqrt{R^2-B^2+\frac{1}{B^2}} - \frac{R}{B^2}\right) + RB^2 - \sqrt{R^2-B^2+\frac{1}{B^2}}}, \quad (37)$$

and

$$\hat{C_2} = \frac{1}{R} - \frac{\epsilon-1}{R}\hat{C_1}\sqrt{R^2-B^2+\frac{1}{B^2}}. \quad (38)$$

Equation (36), therefore, gives the total upward force on the droplet in terms of three non-dimensional parameters Bo_e, Bo_g, and R that are identical to that of the conducting case, and one new parameter

$$\epsilon = \frac{\epsilon_1}{\epsilon_2}. \quad (39)$$

7. Results

As before, we set the expression for total force given in Equation (36) equal to zero, which allows us to express Bo_e in terms of B. This gives three different types of behavior based on the values of ϵ. In case one, where $\epsilon \gg 1$, and R is fixed we predict that the droplet behaves similarly to the conducting drop (see Figure 11). In case two, where $\epsilon > 29$ we have that hysteresis or the presents of two stable branches of fixpoints can occur for certain values of R (see Figure 12). Finally, in the third case, where $\epsilon < 29$, a wide range of stable fixed points exists, which extends to just below the electrode (see Figure 13). The existence of two stable branches of fixed points for certain values of ϵ agrees well with numerical results for the three dimensional problem, as Ramos and Castellanos [27] and Wohlhuter and Basaran [26] predict a similar behavior, however, their minimum value for which hysteresis can occur is $\epsilon = 20$ as opposed to our prediction of $\epsilon = 29$ for the two dimensional problem.

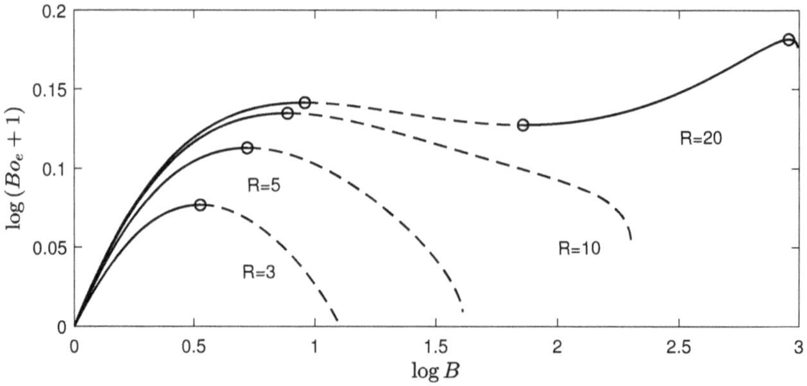

Figure 11. Bo_e as a function of B on a log-log scale for $\epsilon = 78$ (which approximates the relative permittivity of water). We can see that for values of R that are small in comparison to ϵ, the droplet behaves similarly to a conducting droplet.

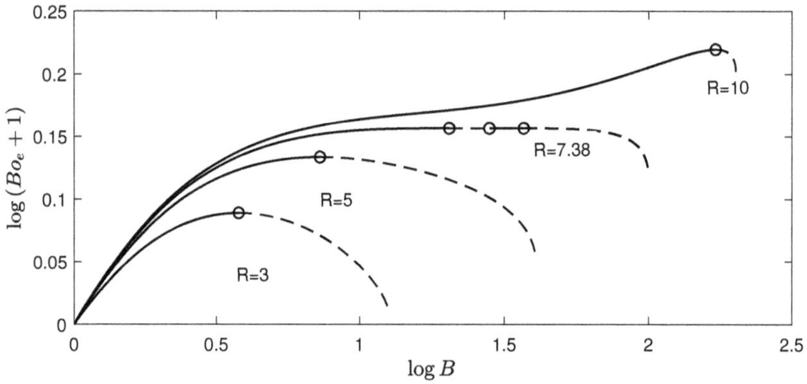

Figure 12. Bo_e as a functions of B on a log-log scale for $\epsilon = 29$, which is close to the minimum value of ϵ for which hysteresis can occur. We see that hysteresis can only happen for a small range of R values.

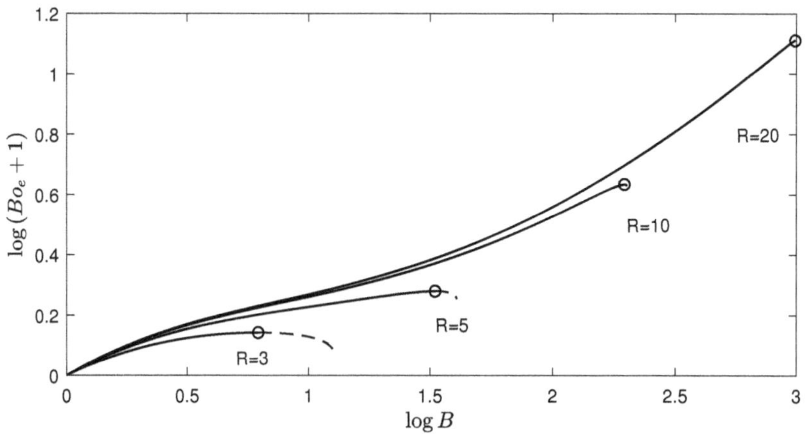

Figure 13. Bo_e as a functions of B on a log-log scale for $\epsilon = 10$, we see that for small values of ϵ a large range of stable fixed points exist, and the two-dimensional droplet can stably approach the electrode.

8. Conclusions

Conformal maps are used to analyze the behavior of a substrate supported two-dimensional droplet that is placed under an electric potential. In both the conducting case and the dielectric case, the behavior of a two-dimensional droplet is similar to that of a three dimensional one. In the case of a conducting drop, we find that the maximum stable high for an ellipse shape droplet is $B = 2.31$, which has a corresponding voltage value of $Bo_e = 0.14$. In the dielectric case, we find that when problem parameters are fixed and permittivity is large, the potential surrounding a dielectric drop approaches the potential that surrounds a conducting drop, and thus a dielectric drop behaves similar to a conducting drop. However, in the intermediate range where the permittivity of the drop is above 29 but also not too large, we have that hysteresis can occur for certain values of electrode separation. Finally, when the permittivity of the drop is below 29 we have that the behavior of a dielectric drop is far different than that of a conducting drop, as a wide range of stable fixed points that rise to just below the electrode exists. In the future, we hope to apply this conformal map approach to address droplets with varying contact angles, and general droplets with interfaces that are not restricted to an ellipse.

Author Contributions: Formal analysis, P.Z.; methodology, S.A.; supervision, S.A.; visualization, P.Z.; writing—original draft, P.Z.; writing—review and editing, S.A.

Funding: This research received no external funding.

Acknowledgments: We acknowledge the partial support by the Petroleum Research Fund PRF-59641-ND and the NJIT Provost's Summer Research Award.

Conflicts of Interest: The authors declare no conflict of interest.

References

1. Jiang, H.; Tan, H. One dimensional Model for droplet ejection process in inkjet devices. *Fluids* **2018**, *3*, 28. [CrossRef]
2. Bos, A.V.; Meulen, M.V.; Driessen, T.; Berg, M.; Reinten, H.; Wijshoff, H.; Versluis, M.; Lohse, D. Velocity profile inside piezoacoustic inkjet droplets in flight: Comparison between experiment and numerical simulation. *Phys. Rev. Appl.* **2014**, *1*, 014004.
3. Eggers, J.; Villermaux, E. Physics of liquid jets. *Rep. Prog. Phys.* **2008**, *71*, 036601. [CrossRef]
4. Gaskell, S.J. Electrospray: Principles and practice. *J. Mass Spectrom.* **1997**, *32*, 677–688. [CrossRef]
5. Wilm, M.; Shevchenko, A.; Houthaeve, T.; Breit, S.; Schweigerer, L.; Fotsis, T.; Mann, M. Femtomole sequencing of proteins from polyacrylamide gels by nano electrospray mass spectrometry. *Nature* **1996**, *379*, 466–469.
6. Yarin, A.L.; Koombhongse, S.; Reneker, D.H. Taylor cone and jetting from liquid droplets in electrospinning of nanofibers. *J. Appl. Phys.* **2001**, *90*, 4836–4846. [CrossRef]
7. Bhardwaj, N.; Kundu, S.C. Electrospinning: A fascinating fiber fabrication technique. *Biotechnol. Adv.* **2010**, *28*, 325–347. [CrossRef]
8. Gierak, J. Focused ion beam technology and ultimate applications. *Semicond. Sci. Tech.* **2009**, *24*, 043001. [CrossRef]
9. Matsui, S.; Ochiai, Y. Focused ion beam applications to solid state devices. *Nanotechnology* **1996**, *7*, 247–258. [CrossRef]
10. Xie, J.; Lim, L.K.; Phua, Y.; Hua, J.; Wang, C. Electrohydrodynamic atomization for biodegradable polymeric particle production. *J. Colloid Interface Sci.* **2006**, *302*, 103–112. [CrossRef]
11. Taylor, G.I. Disintegration of water drops in an electric field. *Proc. R. Soc. A* **1964**, *280*, 383–397.
12. Wilson, C.T.R.; Taylor, G.I. The bursting of soap-bubbles in a uniform electric field. *Math. Proc. Camb. Philos. Soc.* **1925**, *22*, 728–730. [CrossRef]
13. Cheng, K.J.; Chaddock, J.B. Deformation and stability of drops and bubbles in an electric field. *Phys. Lett. A* **1984**, *106*, 51–53. [CrossRef]
14. Cheng, K.J.; Miksis, M.J. Shape and stability of a drop on a conducting plane in an electric Field. *PhysicoChem. Hydrodynam.* **1989**, *11*, 9–20.
15. Miksis, M.J. Shape of a drop in an eletric field. *Phys. Fluids* **1981**, *24*, 1967–1972. [CrossRef]

16. Taylor, G.I.; McEwan, A.D. The stability of a horizontal fluid interface in a vertical electric field. *J. Fluid Mech.* **1965**, *22*, 1–15. [CrossRef]
17. Corson, L.T.; Tsakonas, C.; Duffy, B.R.; Mottram, N.J.; Sage, I.C.; Brown, C.V.; Wilson, S.K. Deformation of a nearly hemispherical conducting drop due to an electric field: Theory and experiment. *Phys. Fluids* **2014**, *26*, 122106. [CrossRef]
18. Tsakonas, C.; Corson, L.T.; Sage, I.C.; Brown, C.V. Electric field induced deformation of hemispherical sessile droplets of ionic liquid. *J. Electrostat.* **2014**, *72*, 437–440. [CrossRef]
19. Ganan-Calvo, A.M. On the theory of electrohydrodynamically driven capillary jets. *J. Fluid Mech.* **1997**, *335*, 165–188. [CrossRef]
20. Lastow, O.; Balachandran, W. Numerical simulation of electrohydrodynamic (EHD) atomization. *J. Electrostat.* **2006**, *64*, 850–859. [CrossRef]
21. Lauricella, M.; Melchionna, S.; Montessori, A.; Pisignano, D.; Pontrelli, G.; Succi, S. Entropic lattice boltzmann model for charged leaky dielectric multiphase fluids in electrified jets. *Phys. Rev. E* **2018**, *97*, 033308. [CrossRef]
22. Elele, E.O.; Shen, Y.; Pettit, D.R.; Khusid, B. Detection of a dynamic cone-shaped meniscus on the surface of fluids in electric fields. *Phys. Rev. Lett.* **2015**, *114*, 054501. [CrossRef]
23. Macky, W.A. Some investigations on the deformation and breaking of water drops in strong electric fields. *Proc. R. Soc. A* **1931**, *133*, 565–587. [CrossRef]
24. Castellanos, A.; Gonzalez, A. Nonlinear electrohydrodynamics of free surfaces. *IEEE Trans. Dielectr. Electr. Insul.* **1998**, *5*, 334–343. [CrossRef]
25. Hua, J.; Lim, L.K.; Wang, C.H. Numerical simulation of deformation/motion of a drop suspended in viscous liquids under influence of steady electric fields. *Phys. Fluids* **2008**, *20*, 113302. [CrossRef]
26. Wohlhuter, F.K.; Basaran, O.A. Shapes and stability of pendant and sessile dielectric drops in an electric field. *J. Fluid Mech.* **1992**, *235*, 481–510. [CrossRef]
27. Ramos, A.; Castellanos, A. Equilibrium shapes and bifurcation of captive dielectric drops subjected to electric fields. *J. Electrostat.* **1994**, *33*, 61–86. [CrossRef]
28. Inculet, I.I.; Kromann, R. Breakup of large water droplets by electric fields. *IEEE Trans. Ind. Appl.* **1989**, *25*, 945–948. [CrossRef]

MDPI
St. Alban-Anlage 66
4052 Basel
Switzerland
Tel. +41 61 683 77 34
Fax +41 61 302 89 18
www.mdpi.com

Fluids Editorial Office
E-mail: fluids@mdpi.com
www.mdpi.com/journal/fluids

www.ingramcontent.com/pod-product-compliance
Lightning Source LLC
LaVergne TN
LVHW071956080526
838202LV00064B/6766
* 9 7 8 3 0 3 9 2 8 2 9 6 8 *